5/20

P9-DTC-024

"Engrossing . . . Crosby, a journalist, profiles the outbreak as it rips through Memphis, the city hardest hit. A first-rate medical detective drama . . . It is good to be reminded of the occasional nobility of the human spirit." —*The New York Times Book Review*

"A fascinating book about yellow fever, its unspeakable horrors and the uncommon valor that four doctors displayed in their quest to solve a devastating medical mystery." —*The Tennessean*

"A forceful narrative of a disease's ravages and the quest to find its cause and cure. Crosby is particularly good at evoking the horrific conditions in Memphis, 'a city of corpses' and . . . also relates arresting tales of heroism." —*Publishers Weekly*

"Seamlessly blends history and science to tell us how yellow fever haunted the nation—and why, if we're not extremely vigilant, it will haunt us again." —Hampton Sides, author of *Blood and Thunder*

"Masterful . . . Crosby uses rich detail and a stunning cast of characters to bring to vivid life the devastating yellow fever epidemic of 1878." —Candice Millard, author of *The River of Doubt*

"Meticulous research and adroit storytelling . . . After a few chapters of *The American Plague,* I had to take an aspirin and lie down—and that is a tribute to the power of Molly Crosby's memorable evocation of a terrible time."

—Robert M. Poole, author of *Explorers House*

THE
AMERICAN PLAGUE

The Untold Story of Yellow Fever,
The Epidemic That Shaped Our History

Molly Caldwell Crosby

BERKLEY BOOKS, NEW YORK

THE BERKLEY PUBLISHING GROUP
Published by the Penguin Group
Penguin Group (USA) Inc.
375 Hudson Street, New York, New York 10014, USA
Penguin Group (Canada), 90 Eglinton Avenue East, Suite 700, Toronto, Ontario M4P 2Y3, Canada
(a division of Pearson Penguin Canada Inc.)
Penguin Books Ltd., 80 Strand, London WC2R 0RL, England
Penguin Group Ireland, 25 St. Stephen's Green, Dublin 2, Ireland (a division of Penguin Books Ltd.)
Penguin Group (Australia), 250 Camberwell Road, Camberwell, Victoria 3124, Australia
(a division of Pearson Australia Group Pty. Ltd.)
Penguin Books India Pvt. Ltd., 11 Community Centre, Panchsheel Park, New Delhi—110 017, India
Penguin Group (NZ), 67 Apollo Drive, Rosedale, North Shore 0745, Auckland, New Zealand
(a division of Pearson New Zealand Ltd.)
Penguin Books (South Africa) (Pty.) Ltd., 24 Sturdee Avenue, Rosebank, Johannesburg 2196,
South Africa

Penguin Books Ltd., Registered Offices: 80 Strand, London WC2R 0RL, England

While the author has made every effort to provide accurate telephone numbers and Internet addresses at the time of publication, neither the publisher nor the author assumes any responsibility for errors, or for changes that occur after publication. Further, publisher does not have any control over and does not assume any responsibility for author or third-party websites or their content.

PRINTING HISTORY
Berkley hardcover edition: November 2006
Berkley trade paperback edition: September 2007

Berkley trade paperback ISBN: 978-0-425-21775-7

The Library of Congress has catalogued the Berkley hardcover edition as follows:

Crosby, Molly Caldwell.
    The American plague : the untold story of yellow fever, the epidemic that shaped our history /
Molly Caldwell Crosby.
      p.   cm.
    Includes bibliographical references and index.
    ISBN 0-425-21202-5 (alk. paper)
      1. Yellow fever—History.   2. Yellow fever—Tennessee—Memphis—History.   I. Title.

  RC211.T3C76   2006
  614.5'41—dc22

                                                                              2006050497

PRINTED IN THE UNITED STATES OF AMERICA

10   9   8   7   6   5   4   3   2   1

CONTENTS

Author's Note

This is a work of nonfiction: Any italicized or quoted statements are taken directly from letters, diaries, articles, books, or actual dialogue. The greatest challenge in writing this book was that the people involved, though truly heroic, were not famous. Even Walter Reed, best known of all the principal characters, has only a handful of biographies to his name, most of which are out of print. To write this story, I relied heavily on personal letters and diaries for character development; the rest was filled in with historical information from the time period and newspaper clippings. In Memphis, I pored through old photographs, newspapers and family papers to re-create the city in 1878. I visited numerous libraries for historical collections pertaining to this story, including the New York Academy of Medicine, which now houses Jesse Lazear's original logbook—it was lost for fifty years before someone found it in a trash can and retrieved it. And I traveled to Havana, Cuba, to visit and photograph the original site of Walter Reed's yellow fever experiments.

I am greatly indebted to Philip S. Hench, a Nobel Prize–winning scientist, whose personal hobby was acquiring and interpreting massive amounts of information on Walter Reed and his Yellow Fever Commission for a book. Hench died before he was able to write that book, and his collection is now held at the University of Virginia; mine is but one of many other books that have sprouted from his years of research and insight.

The scope of this disease and its effect on this country is vast. It was a plague intrinsically tied to the worst and best in humanity, brought on by the mistreatment of others and conquered only by selfless sacrifice. In this book, I hoped to give a poignant portrayal of yellow fever by narrowing the focus to one town, a Southern city that would rise from the ashes, and a handful of doctors, one of whom would rise in the ranks of our country's history. Yet their stories are the stories of dozens of other places and thousands of other people.

Nothing is an accident. Fever grows in the secret places of our hearts, planted there when one of us decided to sell one of us to another.

—JOHN EDGAR WIDEMAN, *Fever*

Memphis, 1870s

The Infected District is outlined
- A. Mississippi River
- B. Wolf River outflow
- C. Focal point of outbreak (Front Street)
- D. Happy Hollow
- E. Pinch District
- F. Gayoso Bayou
- G. Court Square
- H. Board of Health
- I. St. Mary's Episcopal Church
- J. Victorian Village
- K. Peabody Hotel
- L. Cotton Row
- M. Site for Customs House
  (constructed in 1876)

Courtesy of the Library of Congress

MISSISSIPPI RIVER

PROLOGUE

A House Boarded Shut

The flies had been swarming around the house for days. As he walked the exterior, he tried to peer into the boarded windows where flies crawled through starbursts of broken glass. Shielding his eyes with cupped hands, he could not see anything through the black splinters of darkness but a gray, dim light. He had no name, at least not one that survived in the family records; he was an old slave who continued to live with the Angevine family as a servant long after he had been legally freed. The family owned a 4,000-acre farm outside of Grenada, Mississippi, where Mrs. Angevine had been born and raised before marrying a New York attorney and moving to Memphis. A graduate of Harvard Law, Mr. Angevine worked in the Memphis offices of Harris, McKissick and Turley. When the Civil War broke out, he fought for the South against his brothers fighting for the North.

The family left Memphis and boarded themselves inside of

the plantation house when the 1878 yellow fever epidemic struck. The measure may have seemed drastic to others, but Mr. Angevine understood the toll of yellow fever better than most; his wife had died of the fever the previous summer.

As far as the servant could see, the front door was locked, and the flies seemed to have the only access to the house. He pried open the shutters and broke the glass, letting loose a plume of repugnant air. In the stale, dark rooms he saw the corpses of the Angevine family, many in advanced stages of decomposition. Even in the darkness, he could see their yellow skin, the color of unpolished brass. Mary Louisa, the eldest, had been the first to go, and five others had followed. Mr. Angevine lay dead among his children.

No one can really imagine those final days in the fever-ridden house. The fever attacked each person in the Angevine family, one after the other, until none were well enough to help the others. It hit suddenly in the form of a piercing headache and painful sensitivity to light, like looking into a white sun. At that point, the patient could still hope that it was not yellow fever, maybe just a headache from the heat. But the pain worsened, crippling movement and burning the skin. The fever rose to 104, maybe 105 degrees, and bones felt as though they had been cracked. The kidneys stopped functioning, poisoning the body. Abdominal cramps began in the final days of illness as the patient vomited black blood brought on by internal hemorrhaging. The victim became a palate of hideous color: Red blood ran from the gums, eyes and nose. The tongue swelled, turning purple. Black vomit roiled. And the skin grew a deep gold, the whites of the eyes turning brilliant yellow.

* * *

The servant climbed through the window and made his way through the rooms where other servants and guests had died. Finally, he found a body not yet rotting; it was the youngest daughter, nine-year-old Lena. He knelt down, brushing away flies and maggots. Lifting her weightless frame, he carried her out of the family mansion that had now become a tomb, into the fresh air. He placed her body in a nearby house, resting a piece of raw bacon across her lips, and watched as Lena began to suck on the first bite of food she had had in days. It was then that Lena, more dead than alive, began to make her way back.

In the coming weeks, she would recover and tell of the horrors trapped inside of the house. Men from the countryside came into the city, robbing the dead and stepping over the bodies of the dying. "My own father, while ill with the fever, was choked, robbed and left alone to die: I was too ill to even cry out for help, but witnessed the entire affair."

With no surviving family, Lena went to live with her grandparents in Memphis to attend St. Mary's School. After graduation, and against the wishes of her grandparents, she enrolled in a nursing school at the Maury and Mitchell Infirmary, where she worked under the tutelage of Dr. Robert Wood Mitchell.

The remainder of Lena Angevine's time in the South was a patchwork of nursing, ambition and marriage to a St. Louis man by the name of E. C. Warner. Warner, whose name Lena Angevine would take from that point forward, died of a heart attack just four months after the marriage. At least that was the generally accepted story; family rumors also alluded to a divorce. As it was far better to be a widow than a divorcée in the 1890s, Lena Warner would naturally opt for a dead husband over an estranged one.

Lena Angevine Warner's experience with yellow fever would continue to sear her, and in 1898, with the outbreak of the Spanish-American War, she answered an ad in the newspaper. The surgeon general of the United States Army was looking for nurses immune to yellow fever. Warner would receive fifty dollars per month, and in 1900, she was stationed in Cuba as chief nurse under a doctor of some distinction, Major Walter Reed.

PART ONE

The American Plague

Plague: A widespread affliction or calamity, especially one seen as divine retribution.

—*The American Heritage Dictionary*

The rain came in West Africa. A massive wind blew in from the Atlantic coast bringing the deluge of water known as the southwest monsoon. It swelled the Niger and Benue rivers; it spilled into the braided streams of the Niger Delta; it filled the floodplains and swamplands of southeastern Nigeria. It purpled the sky and saturated the country.

Towering oil palms dripped water from their feathered branches, and the brush-marked trunks of rubber trees were streaked with rain. The broad, bush-topped cocoa trees moistened. And the forests of Nigeria grew heavy and humid. Water is nourishing, and it enriched the plant life emerging from West Africa's dry season. It also nourished something else—dry, oval-shaped eggs clinging to life inside the hollows of trees. Once the rain fell, those eggs grew, and soon, mosquitoes hatched. In the natural world, a string of events had been set into motion.

* * *

The rains falling in Africa did not deter men from entering the forest and felling trees for timber. As the trees fell, their canopy of dense green fell with them, often bringing a swarm of gnats and mosquitoes with it. Some of those mosquitoes would bite.

The native Africans who worked the forests noticed an eerie silence in the trees. Usually alive with the piercing sound of birds, the hum of insects and the calls of monkeys, tree canopies in some areas were still, a haunting contrast to the living, breathing rain forest—a sign that something was not right in the ecosystem. The monkeys had grown ill, their shrill chatter quieted. Unknown to the men, the rain forest, teeming with smells, sounds, color and life, was also home to something much smaller. Microscopic. A tiny, thriving life-form.

No one knows for sure how the yellow fever virus first came into existence. No records of it in early history exist, nor is it among the biblical plagues. But then, how does any new life emerge? There is a creation and a birth and eventually a discovery in the dark forests of Africa.

A virus is one of the smallest beings in evolution's survival of the fittest, mutating and coalescing in order to thrive, its ultimate goal being epidemic. Viruses affect nearly every life-form on earth from flora to fauna, but a virus in its own right is not actually alive—it only becomes alive by possessing something living. The virus seeks out a healthy cell, overtakes it, impregnating it, forcing the body's cells to produce thousands of the new offspring. This rapacious battle will eventually allow the virus, something as small as one-ten-thousandth of a millimeter, to conquer something the size of a human.

It is uncertain whether viruses evolved from a single cell, becoming more complex, or whether they *devolved* into something simpler, more efficient, gracefully infectious. Either way, a virus is an evolutionary masterpiece; since it does not have the ability to have sex or reproduce on its own, it must constantly change, adapting to other life-forms—from something as small as bacteria to something as large as mammals. Taking it a step further, once the virus has mastered something like a mosquito in the case of yellow fever or a bird in the case of influenza, it may spread to other species. There, it adapts again and again until there is a seamless transition between certain species—perhaps a monkey to a mosquito to a man.

Once inside the bloodstream, a virus is programmed with elegantly simple genetic material, its DNA or RNA, to produce certain symptoms that will spread it further. In the common cold, sneezing and a runny nose spread the virus. In smallpox, open sores on the skin act as the vehicle for infection. With influenza, coughing expels the virus into the air. In the case of HIV, the virus uses one of the most basic functions in human life—reproduction—to spread through the population of men, women and children.

Yellow fever is what is known as a flavivirus, a group of viruses spread by mosquitoes that includes West Nile, dengue and Japanese encephalitis. As a virus, yellow fever is not one of the stronger ones. It cannot live outside of the body for more than a few hours. It does not spread through the air or by touch. It does not mutate as easily as some viruses. In fact, its most telling symptom—fever—is really just the body's own attempt to kill the virus. What makes yellow fever unique is its choice of vector. What the virus lacks in evolutionary prowess, the mosquito makes up for.

In the African rain forest, mosquitoes carry the virus, this finely evolved life-form looking to conquer. The mosquito feeds off of the monkeys in the tree canopy, and a small epidemic erupts, foreshadowing the larger one to follow.

Of course, none of this would be known until well into the twentieth century. For the Africans and Europeans, and later for the Americans, a virus would remain an invisible, unknown entity. No one would even know how yellow fever spread from one person to the next until 1900.

As the men made their way back to the Niger Delta and the coast of West Africa where the timber would be sold for ships carrying palm oil, ivory, salt, gold and slaves, they might run a mild fever or feel lethargic, but it was nothing compared to what the white Europeans would feel in the coming weeks. Through this cycle of men entering the forests, mosquitoes biting men and the virus spreading among small tribal villages, most native Africans had encountered the yellow fever virus at one time or another. They acquired immunity to it, and the virus began to run out of the kindling that kept the flames of fever alive. When white Europeans landed on the coast of West Africa, it was like a fresh burst of oxygen in a waning fire.

The Nigerian coast had been booming with the slave trade since the fifteenth century, providing the Middle Passage across the Atlantic with 30 percent of its human exports. As the ships of the Portuguese, Spanish, Dutch and English pitched in the ocean swells along the coast, they waited for the cargo making its way from the interior down the river delta to the coast.

The port towns were filled with the smells of Africa: sweet oil from the palms, spices, yams and dates sold at local markets. There was fire smoke and the poignant scent of rainwater mixed with human sweat where the next shipment of slaves sat chained to one another in thatched sheds, waiting to board the ship. Fearful of tropical diseases, the Europeans might even taste the sweat of slaves to try to determine if he or she carried disease. The smells would worsen in the next few months at sea when hot air, sweat and human excrement would be trapped beneath deck with the slaves. When the ships encountered squalls, and the sea and sky would join, the tumbling ship would induce vomit to add to the amalgam of human smells.

As the ship traversed the waters of the Atlantic, the virus made its way through the bloodstream of the passengers as succinctly as it had made its way through the rivers of West Africa to the coast. In the blood, yellow fever looks something like a fuzzy snowflake, but it is actually round with twenty smooth sides that protect the virus's single strand of RNA at the center. The coating of the virus is made up of proteins, and human cells are attracted to those proteins—the virus doesn't need to look for healthy cells; they look for it. Once the two make contact in the bloodstream, a process with a technical name known as receptor-mediated endocytosis begins—sort of a molecular Trojan horse. The healthy cell eventually enfolds the virus, taking it in and closing the door behind it. Once inside, the virus hijacks the cell and its basic machinery, using the cell's internal makeup to replicate the viral proteins and RNA, until the new particles burst through the cell. A body that was once filled with healthy cells is now filling up with cells carrying the yellow fever virus. That is why a virus cannot be treated by antibiotics; human cells give it refuge, and anything that could destroy the virus might also destroy the cell.

Soon, the filth, lack of nutrition, dehydration and rapid-fire spread of disease turned the transatlantic journey into a death voyage, with bodies being tossed overboard in the wake of the ship. In fact, slave ships were often trailed by sharks, which quickly learned that the vessels served as a source of food.

It was through this journey from the interior of West Africa, down the Niger and Benue rivers, to the coast, onto ships and into the blood of Europeans that yellow fever first made its way from the Old World to the new one.

Yellow fever, more than any other disease, would seem conjured by God and divinely directed. When the slave trade first began, every European country that profited from the purchase and sale of Africans would soon see a yellow fever epidemic: the United Kingdom, France, Germany, the Netherlands, Spain, Portugal. Though Asia had the ideal climate and the right mosquito, it has never had an epidemic of yellow fever. It also never participated in the African slave trade.

As the European powers crossed the Atlantic to establish West Indian colonies, which quickly became horrific holding pens for slaves, yellow fever settled its roots in the western hemisphere and proliferated. The first epidemic on this side of the world occurred in 1648. After that, the slave trade increased fivefold in the West Indies. And by 1702, as the trade of flesh spread to North America, yellow fever blossomed on the continent. From 1700 to 1750, the slave population in America doubled and then doubled again. As each slave ship arrived into the ports of the New World, bringing over ten million slaves to this hemisphere, yellow fever made a giant, evolutionary leap. It adapted. It spread. As one historian put it, "When the disease invaded the Atlantic and Gulf States, it struck with a force more powerful than the one which bombed Pearl Harbor more than two centuries later."

Yellow fever became the most dreaded disease in North America for two hundred years. It did not kill in numbers as high as some of its contemporaries like cholera or smallpox, and it was not contagious; yet it created a panic and fear few other diseases, ancient or contemporary, can elicit.

During its tenure in this country, yellow fever would inflict 500,000 casualties and 100,000 deaths. The fever would stretch the length of North America, afflicting Massachusetts, Rhode Island, New Hampshire, Connecticut, New Jersey, Pennsylvania, New York, Delaware, Maryland, Illinois, Missouri, Ohio, Kentucky, Virginia, North Carolina, South Carolina, Georgia, Alabama, Tennessee, Mississippi, Arkansas, Louisiana, Florida and Texas.

The U.S. capital would move from Philadelphia to Washington, D.C., after a devastating yellow fever epidemic in 1793. Alexander Hamilton suffered the fever, while George Washington, John Adams and Thomas Jefferson fled the city; the United States government was paralyzed.

In New York, Greenwich Village would become known as "the Village" because it was the safe haven outside of the city during yellow fever epidemics.

Napoleon would abandon his conquests in North America after losing 23,000 of his troops to yellow fever in the colony of Haiti. He made a hasty and fearful retreat from this pestilent hemisphere, selling his large Louisiana holdings for cheap to Thomas Jefferson.

During the Civil War, yellow fever would serve as one of this country's first forms of biological warfare. And the Spanish-American War, at the close of the nineteenth century, would be fought more against this fever than against the Spanish.

For the first century of its siege in the United States, yellow fever marked for destruction the heavily populated, northern port cities of Boston, New York and Philadelphia. Then, in 1807, the Atlantic slave trade was abolished, and the fever suddenly retreated from the North. By 1850, no other epidemics of yellow

fever would occur in those major cities. As the North weaned itself from the slave trade, its southern counterpart absorbed the
slave labor and the accompanying yellow fever. In the South,
where slavery became deeply entrenched, yellow fever found its
lifeblood.

The 1878 yellow fever epidemic, the worst in history, started with
the rains in West Africa. February, the wet season, arrived, the
mosquitoes hatched, the monkeys grew ill, the loggers stared up
at the silent tree canopy. This time, the ships moored off the coast
would not carry slaves across the Atlantic; they would carry ivory,
gold, copper, salt—and mosquitoes.

But this year would be different for two reasons: Nature had
afforded the virus with the perfect environment. An El Niño cycle
turned the American South that winter into a tropical region with
warm temperatures and rainfall 150 percent above normal. Insects, usually deterred by the winter freeze, proliferated. The
significance of the weather phenomenon meant nothing to
nineteenth-century observers, but 100 years later, scientists
would link El Niño to most major outbreaks of yellow fever. As
southerners cut hyacinth blooms in January and waded through
waterlogged streets, they complained about the number of mosquitoes beginning to swarm.

American progress was the virus's other ally. A great influx of
immigrants—Irish, German, eastern European—had been mi-
grating south since the Civil War. Just like the white Europeans
descending upon Nigeria and other parts of West Africa, they
served as fuel for a fever fire, a fresh source of nonimmune blood
for the virus.

Transportation had paved the way for these immigrants.
Trains connected every corner of America for the first time—east

to west, north to south. And paddleboats and steamers snaked their way north from the Gulf of Mexico up the Mississippi River. At the center of this web sat a city 400 miles inland from the Gulf, ready to take its place as one of the largest, most successful cities in the South.

Memphis, Tennessee, was poised for greatness in 1878. By the end of that year, it would suffer losses greater than the Chicago fire, San Francisco earthquake and Johnstown flood combined. The devastation to the Mississippi Valley would cost over $350 million by today's standards. And the U.S. government would create the National Board of Health, which would report: "To no other great nation of the earth is yellow fever so calamitous as to the United States of America."

As the southwest monsoon pelted the Niger Delta in February 1878, hatching mosquito eggs and giving birth to a virus, people on the other side of the world could not have known what awaited them. In Memphis, Tennessee, their attention was turned not toward disease or death, but just the opposite: a carnival.

PART TWO

Memphis, 1878

And now was acknowledged the presence of the Red Death. He had come like a thief in the night. And one by one dropped the revelers in the blood-bedewed halls of their revel.

—EDGAR ALLAN POE,
"The Masque of the Red Death"

CHAPTER 1

Carnival

The bell sounded.

A servant wearing a white jacket, with all the trimmings of formality, stood outside the door, a gilded envelope in his gloved hand. It was the most coveted invitation of the year.

The envelope was exquisite, large and square, with golden calligraphy. Inside, it took the shape of a scroll on powder-blue parchment with a regal crown framing the top where CARNIVAL: MEMPHIS MARDI GRAS was engraved. Fanning out of an Egyptian pyramid, the secret order of the Memphi and Ulks invited *you and your household to attend his pageants March 4 and 5, 1878.*

Over 10,000 people would answer this invitation to Memphis including, one year, the president of the United States. As many as 40,000 revelers would stand shoulder to shoulder along the downtown streets of Memphis. *Harper's* would reserve front-page coverage, sending their best illustrator. The glitter and glamour of the event was known across the country, and it was widely whispered that New Orleans had sent scouts to Memphis to study the parade.

And so began Carnival.

Memphis had been chosen as a bluff city, literally poised on the precipice of the American South and an immense, new frontier. The only thing separating the two was the treacherous Mississippi River, a huge gash in the American landscape. Murky, ochre-colored and unpredictable, the river pushed against levees, dividing the river town from the dense forests and willow thickets of Arkansas. It was the last stop for the likes of David Crockett and Sam Houston making their way to Texas, a place for flatboats to purchase firearms before heading west, and it was the point of entry back into the civilized world of the Old South. But it wasn't just the topography that gave Memphis a startling sense of contrast; it was the people. All classes of society, all colors of skin, all manner of accents migrated to a fault line carved between the past and the future, the Old South and the new frontier.

As the latticework of American transportation and expansion spread westward in the 1850s, Memphis remained at the crossroads; steamboats joined the city to a massive trade line between the Gulf of Mexico through the Ohio Valley, and railroads connected it to the burgeoning ports of Charleston and New Orleans. Surrounded by rural states and plantations, Memphis became a hub: the largest inland cotton market, at its peak, handling 360,000 bales of cotton per year. As the bluff city sloped toward the Mississippi River, levees and thoroughfares piled high with crates of tufted white, Memphis looked like a town literally built upon cotton. But cotton was not the only business booming. At the center of a vast web of plantations, railroad lines and port towns, Memphis profited from the slave market as well. The Bolton, Dickens & Co. held what could only be called "yard sales" for slaves, while Hill, Byrd & Sons and Nathan Bedford Forrest opened slave showrooms on Adams Street. Said to have kept his business fair and his slave pens clean, Forrest prospered as one of the South's largest slave traders, selling up to 1,000 slaves per year. Forrest, a vehement Confederate, took up arms during the Civil War and, in spite of near illiteracy, rose in rank to become one of the greatest military tacticians in American history. His surprise attacks on northern troops in Memphis would later inspire the German blitzkrieg. Long after the Civil War, Forrest's name would live in infamy for founding the Ku Klux Klan.

Even as northern troops marched through Memphis in the 1860s, they recognized it as a center for trade and transportation, contraband or otherwise. It was spared Sherman's flames, and a strange coexistence emerged between the occupying army and its Confederate residents. The North looked the other way as illegal southern shipments slipped through northern quarantine. As a result, despite four years of Civil War, Memphis business and shipping never ceased, and a number of those northern troops never left. The Civil War put an end to slavery once and for all, but the slave trade would have a lasting legacy not yet realized.

By 1870, Memphis's population of 40,000 was almost double that of Nashville and Atlanta, ranking it second only to New Orleans as the largest city in the South. As its population grew, so did its diversity. It was a city built upon clashes, of river against land and people against people.

Despite the relative good fortune in Memphis, the country lagged under national debt in the 1870s. The Panic of 1873 ushered in an economic depression the likes of which had never been seen

before, and the South suffered most of all. The war had destroyed
the vast farmlands and plantations that carried the financial success of the entire region. Even if Memphis was equipped and
ready to ship cotton north, none was arriving from the south. As
jobs grew scarce on farms and in small towns, poor families
flocked to nearby cities, and the newly termed *tramps* moved
freely along the railroads.

Soon, Memphis swelled with the underclasses. A freedman's
camp established during northern occupation propelled the black
population in Memphis from 3,000 to over 15,000, nearly 40 percent of the city's total makeup. Blacks constituted only one fac-
tion of a largely disparate and unhappy population. Irish and
German immigrants struggled to free themselves from the mire
of postwar poverty and racial politics. Yanks continued to live in
the South to facilitate reconstruction. And the upper echelons,
white aristocrats of a bygone era, were depressed by a river town
less southern and far more western, full of filth, violence and
rough river folk. City legislators spent money extravagantly and
foolishly. Indifference from the upper classes broadened the divide, and the by-product of such disregard was filth. In all, it created an atmosphere of dissatisfaction in every sphere of Memphis
life, and the resulting discontent became an anchor heaving
against the city's progress. By the 1870s, Memphis was still very
much a city balanced on the edge of the Mississippi River and its
future, both a coarse river town and an extension of the dying
South.

General Colton Greene had an idea. With an isosceles nose and a
walrus moustache, Greene had a profile like Joseph Stalin. He was
a lifelong bachelor, president of the State Savings Bank, owner of
an insurance company, future founder of the Tennessee Club, a

man who liked to make things happen. He had traveled the world and even held a permit for admittance to private rooms of the Vatican. Greene arrived in Memphis, emerging from the war a hero with a little-known past and a hazy air of nobility. He had been on the frontlines of every battle and wounded three different times; Colton Greene liked to face a challenge head-on.

Greene enlisted the help of Memphis's most influential business and political leaders. This was, after all, the decade in which anything could happen. Thomas Edison had established his Electric Light Company. Johns Hopkins University, the first European-style medical institute, had opened. Alexander Graham Bell patented the telephone. George Eastman was developing the first Kodak camera. Great men were doing great things.

Greene and others like him believed Memphis, for all its flaws, had a very bright future. Already four train lines rumbled through the city, and like the birthmark for any progressive place, the skyline ballooned with smoke from steam-powered mills. Cotton presses bellowed all hours of the day and night. Over fifteen hundred buildings were under construction, and mule cars screeched along metal tracks. The new Peabody Hotel, where whole plantations had been won or lost over a hand of cards, boasted a chef all the way from swanky Saratoga Springs in the Northeast. The grand duke of Russia had made a recent visit to the city on the bluffs. Even the former president of the Confederacy had chosen to make Memphis his home. And people strolling down Main Street literally walked among the crates of cotton, strands of the white gold catching on their clothes and hanging like a talisman. All Memphis needed was a push in the right direction to take its place among cities like St. Louis and Chicago.

There was another reason to attract some positive, national attention to Memphis: disease. Over the last decade, Memphis had earned a reputation as a medical town, in part because the north had used it as a hospital center during the war, but mostly because epidemics had recently rained upon the city. Two yellow fever epidemics, cholera and malaria had given Memphis a reputation as a sickly city and a filthy one. It was unheard of for a city with a population as large as the one in Memphis to have no waterworks—the city still relied entirely on the river and rain cisterns to collect water, and there was no way to remove sewage. While the bluffs afforded a high, beautiful vista of the river, they also sloped back into the heart of the city. Rain streamed down the backside of the bluffs into the town, and further still, into the Gayoso Bayou, which wrapped from the Wolf River in a series of stagnant pools through the entire city, north to south. With every downpour, downtown privies backed up and drained the sewage into the bayou. There was no money or organized method of removing refuse from the bustling city center, so people carted their own garbage to the Gayoso and dumped it. Horse manure and dead animals floated through the pale green scum. Corrupt politics kept city funds depleted, and anything as bland as sanitation or water management was the last thing on the minds of civic leaders. As one historian put it, "The trouble with Memphis was that it simply refused to take the time to make the sometimes painful distinction between prosperity and progress."

To the nation at large, Memphis began to appear as a city of deplorable sanitary conditions and disease. There had to be a way to show the nation that Memphis was not just a stricken city of riverboat gambling and death carts. In 1872, General Colton Greene decided that what the depressed river town needed was an elaborate party. Greene was not the one who originally came up with the idea of Mardi Gras. David P. "Pappy" Hadden holds that distinction, but true to his nature, Greene became the one on the frontlines, and he is credited with the magnificent parades in the following decade.

Greene entreated upon the railroads to lower fares and the local merchants to discount supplies, and then he chose the date: Fat Tuesday, on the eve of Lent. Early March would enable the local cotton farmers and their families the chance to attend before planting season began. As with most other aspects of everyday Victorian life, time, seasons and even society would be directed by farming. Mardi Gras would be the grand finale to the Memphis social season, which began each winter when the harvest was over and the first frost fell, quelling the outbreak of disease.

Greene's handful of leaders called themselves the Mystic Memphi, and their secret society served as the main body of wealth and power for the city. Their names were never revealed, and in fact, their existence never even admitted until decades later. Greene convened clandestine gatherings in a real estate office overlooking Court Square and Second Street. The night a meeting was to take place the newspaper simply published the letters UEUQ, taken from the gates of ancient Memphis, Egypt; only those who knew the meaning need answer the call. Over the next several years, the Memphi, and their younger, rowdier counterparts known as the Ulks, would organize the most lavish Mardi Gras celebrations ever seen.

It was unusually warm as the final preparations for the 1878 Mardi Gras were under way. Clusters of white sprang from the branches of the peach and pear trees, crocuses and daffodils had blossomed in January and long ago dropped their petals. And the Mississippi River gave off the scent of silted water warm with sunshine.

The curious heat wave only enlivened the festivities; planks of fresh pine were piled along Main Street as skeletal bleachers took shape among the buildings. Crates of champagne arrived, and fine wines hidden during Yankee occupation surfaced. Costumes from

Paris were unpacked, and Confederate gray not seen since the end
of the war was pressed. Lowenstein and Brothers advertised silks,
evening brocades and satins, not to mention the accompanying
opera fans, cloaks and kid gloves. This year's Mardi Gras would
even boast the new, brilliant effect of artificial light at night.
Memphis prepared for what would surely be the grandest of all
Mardi Gras celebrations. In 1878, $40,000 — by today's standards,
well over $1 million—would be spent through private funding on
the extravagant parade celebrating the King and Queen of Memphis society.

Not everyone in Memphis would take part in Carnival—after
all, the city may have had 115 saloonkeepers, 18 houses of prostitution and roughly 3,000 dope addicts, but it also had close to two
dozen churches. The temperance supporters opposed the drinking
that accompanied the festivities. Considered a heathen celebration, Mardi Gras was blasted from the pulpits with dire predic-
tions of wrath and doom. Colonel Charles Parsons preached no
such warnings, however. One week before the parade was to begin, Charles Carroll Parsons stood in full uniform before the
Chickasaw Guards, a civilian military corps, as their chaplain. In
fact, the Chickasaw Guards would be among the local corps to
march in the parade.

Parsons was a lean man with a soldier's build. He had carved
cheekbones, fair hair, a handlebar moustache and a tender smile.
His eyes were deep-set and gave the appearance of sincerity, but
there was also something intense in his expression. He was once
described as having a look near fanaticism in his face, a passion for
what he believed to be his calling and duty. Not a single surviving
letter or description describes him as anything other than gentle
and great; and in spite of being a Yankee, Charles Parsons was one
of the most beloved rectors in Memphis.

During the war, Parsons had been a northern officer and a

hero. At the Battle of Perryville in Kentucky, he continued operating a gun, single-handedly, after all of the officers and men in his company had fallen. When the Confederate artillery approached, Charles Parsons held his sword at parade rest and awaited fire. The Confederate colonel, impressed by his courage, ordered his men to hold their fire and allowed Parsons to walk off the battlefield. "That man," the colonel exclaimed, "is too brave to be killed."

After the war, Parsons taught at West Point and served with General George Custer in the western campaigns. Custer, a friend and admirer, tried to persuade Parsons to remain in the military, but Parsons felt a different calling. He soon took his orders as an Episcopal priest from Tennessee Bishop Charles T. Quintard, another veteran of the Battle of Perryville, but one who fought on the other side. Parsons came to Memphis to grieve the loss of his first wife, who died in childbirth, and start anew as rector for Grace Church, where this Union officer now preached to a congregation that included Jefferson Davis and his family.

On that late day in February 1878, in a city filled with the sound of hammers and the scent of lumber, Parsons preached not about heathen celebrations or temperance, but about the character of men: "There will come to each of you a time, I trust far away, when the scourge of affliction may fall heavily upon you . . . wealth, or power, or skill, or even fond affection in the utmost stretch of tenderness, can supply no companion to the soul in its journey through the valley of death." He spoke with the confidence of a soldier who had survived the Civil War, the death of a wife and the loss of a son to scarlet fever; for all intents and purposes, he had been to that valley and returned. Parsons did not know at that moment what lay ahead, that the greatest American urban disaster to date awaited them, that when the fever would finally take him, he would have to read his own last rites.

The room was still as Parsons spoke of measuring a man's spirit and strength against the darkest moments, and then he ended his sermon ". . . I was about to say, God send us such a man. I think it is better that I pray—God make us to be such."

For weeks, the Memphis *Appeal* devoted columns to the upcoming parades, their themes, routes and security. The entire police force would be on duty downtown, and concealed weapons would be prohibited. Public drunkenness would not be tolerated, nor would revelers costumed in such a manner that would "shock the decency of the occasion." But the paper also focused on some important national news. The silver bill before Congress authorizing the minting of the silver dollar had graced the front pages. There were the usual mentions of steamboat disasters, train wrecks and the wearing away of Niagara Falls. On Fat Tuesday itself, the paper even made room to report on the Geographical Society's year in research, which included headlines about "Mr. Edison's wonderful phonograph" and "Mr. Stanley's exploration of the Congo."

On Monday, March 4, 1878, Carnival began. Hundreds of people arrived by steamboat and railroad on Sunday, and on Monday, thousands more. As the steamboats rounded the bend, past the mouth of the Wolf River, the bluffs grew closer. Flat-bottom boats bobbed at the water's edge where bands played music on deck. The smokestacks of steamboats crowded the shore like metal tree trunks sprouting smoky limbs. The Memphis skyline was impressive, highlighted by a long row of white, Greek-revival and Italianate structures: There was the exquisite water-pumping station with soaring windows and water fountains; the magnificent Customs House, still under construction, and designed by the architect of the U.S. Treasury; and along Cotton Row, the Gayoso Hotel with its massive columns, still closed for postwar renovations. As though the bluffs had been tipped in frost, the skyline appeared in cream-colored columns, archways and glass. And in the distance, the steeples of St. Mary's, Grace Church and Calvary towered above. There had been a cloudy start to the day, but now there was sunshine and a turquoise sky as beautiful as a summer day; when the sun set across the river that night, the white skyline was lit by the copper light.

Over 10,000 tourists flooded the streets of downtown. Champagne flowed from the fountain in Court Square and a pyramid of candy, twenty-five feet wide and thirty-five feet high welcomed visitors. Many from the countryside came to the shops, where keepers had dropped prices in anticipation of the crowds. Along Main Street, Lowenstein and Brothers clothing store sprawled across three buildings. Kremer, Herzog and Co., a millinery, sold straw hats, netting, flowers and ribbons. Elsewhere downtown could be found lace mitts, gilt hair ornaments, grosgrain ribbon, lisle-thread hose and fresh French flowers. Tobacco shops carried fine cigars, and liquor stores put out their best whiskey. Lloyd's offered sodas, pure cream and caramels.

Some pedestrians rode the electric streetcars; others visited the Memphis Exposition Building, which had opened in 1873. Its architecture reminiscent of an Indian temple, the Exposition Building flew flags from six towers and another forty flags fluttered along its roofline. Many tourists strolled past Jefferson Davis's home on Court Street, while others walked the streets of Adams and Jefferson to admire the most elegant neighborhood in the city. Here, $100,000 was spent building one home. A neighboring house held modern conveniences like airshafts inside its eighteen-inch walls to circulate cool air. Homes in this neighborhood represented contemporary architecture at its best. What would later be known as the Fontaine House, a French Victorian, boasted tin eaves, terra-cotta lintels, and a five-story tower. Its neoclassical neighbor was adorned with Doric columns, and if people strolling the street were to look into the windows, they would see a gaslit Waterford chandelier. Finally, there was a Greek revival with its fluted columns and lotus-leaf carvings. It belonged to the Confederate general Gideon Johnson Pillow.

As the morning progressed, people made their way back downtown, where the excitement was palpable. At noon on Monday, March 4, cannon blasts announced the arrival of Rex and his queen. Observers remarked that as the king appeared in the distance, "the Mississippi trembled underneath her banks."

Dressed in regal purple, Rex approached Mayor John R. Flippin on the grandstand and when the crowd quieted enough, his voice could be heard: "I do now in the name of the Great Momus, High and Mighty Monarch of Misrule, demand the keys to this, my royal master's loyal city of Memphis." The golden key to the city was handed over, the bands struck up and the parade moved forward into the city.

Nightfall began the procession of the Ulks, followed by a magnificent ball. Their theme was the Romances of Childhood, complete with floats of "Hey, Diddle, Diddle," "Jack and the Beanstalk," "Old King Cole" and "Rip van Winkle," among others. As the parade marched by Adams and Second, the First Baptist Church held a lecture on the temperance movement. When the floats passed, the ministers stood outside signing up new volunteers for their "Red Ribbon brigade." All of this mattered little to the Ulks and their throngs who made their way to the Greenlaw Opera House on the corner of Union.

The Romanesque Greenlaw had seen better days. The four-story opera house had been built with greatness in mind: a ballroom that could accommodate three hundred couples; an opera house with eight-foot-wide doors operated by pulleys and counterweights; an auditorium with fifty-foot ceilings. Its builders had visions of the country's finest musicians, theater and modern lectures. But by 1878, it had become a nickel-and-dime hall whose main lectures were those of the temperance movement. Instead of violin concertos, patrons heard a man play his cornet imitations and listened to a sermon on the temperance cause. That year, Mardi Gras breathed a little life and sophistication back into the Greenlaw.

The Greenlaw had been decorated like a childhood fairyland with evergreens, hundreds of imported flowers and caged canaries. A revolving pyramid, three tiers high, repeated the themes of the parade floats. The newspaper reported that "fans fluttered and diamonds flashed." In fact, the only drawback to the evening seemed to be the number of ladies carrying stylish feather fans, which as they quivered, set off tufts of fleecy clouds all over the room. To the intoxicated revelers, it must have seemed like part of the magic.

The biggest attraction, the parade of the Memphi, was yet to come. By Tuesday morning, March 5, costumed people began lining the streets at 10:00 a.m. The costumes were beautiful as often as they were grotesque, and ladies carried horsewhips for their own protection. Again, it was a clear, sunny day, and by twilight, police on horseback marched through the city pushing back the crowds and reminding storekeepers to extinguish the lights inside. In a dark city filled with anticipation, the light show began. Rockets and fireworks exploded over the bluffs, spelling MEMPHI in the night sky. At the water's edge, flat-bottom boats floated like vessels shuddering with candlelight. Street corners were lit by beacons of blues, reds, greens and golds from burning calcium lights.

Bleachers lined the muddy thoroughfares of Main Street

where Wilkerson's apothecary, Harpman and Brother Cigars, and McLaughlin's grocery stood. On every rooftop men dangled their legs over parapet walls and cornices. Women and men crowded in the windows above Barnaby Furnishings and McClelland Drugs, waving streamers. And hundreds of gas lamps, like globes glowing against the March sky, illuminated the streets.

Long before the parade came into view, the trumpets could be heard. The smell of kerosene-laced smoke drifted from torches. The scent of manure was heavy in the air as 3,000 horsemen made their way through downtown, the horse hooves thumping against hard, packed mud. Fireworks ignited the nighttime sky, adding to the dizzying haze and tinderbox smell.

As the first float came into view the cheers of the crowds turned to a roar. The floats were horse-drawn wagons carrying wooden stages, several stories high, draped in vibrant bunting, tapestries and ribbons. Coils of smoke rose from dampened torches to appear as though the floats hovered on clouds. Memphis prided itself on educating its people through the Mardi Gras celebration and float designs. This year, the Mardi Gras theme was the Myths of the Aryan People, which was a relatively modern idea grafting legends and lore of Asia with Europe. "From identity and language, we know that what now constitutes many and mighty nations there all descended from one common stock," the Memphis *Appeal* wrote as way of explanation. Fifty years later, Adolf Hitler would usurp the term *Aryan* for his own definition of pure, Nordic descendants.

That night, *Blue Danube* and *Golden Slippers* could be heard through open windows as orchestras throughout the city played. Every building or residence with a ballroom hosted a party, but the grand masked ball took place at the theater on Jefferson Avenue.

It was to attend Rex's ball that one needed the coveted, gilded invitation, hand delivered by servants. The men attended in lavish costumes or Confederate uniforms, while the women wore gowns of silver brocade and rich velvets; ornate fans of ostrich feathers, organza and rice paper fluttered in their hands. The invitation marked not only entry to the ball but also access to the innermost circles of Memphis society. Here, owners of those Victorian mansions on Adams gathered, where the Overtons, Toofs, Trezevants, Snowdens and many other oft-mentioned names would celebrate the fortune of their city.

Of course there were problems; the town was $4 million in debt. There was not enough money to remove garbage and refuse from the streets. There were cotton crops to be planted and gambling debts to be paid. But for these two days, both ends of the economic spectrum donned masks and overlooked their discontent. Blacks and whites, immigrants and southern elite, businessmen and boatmen could only see the brilliant parade before them, the electrifying colors, the baskets of champagne, the intoxicating sense of well-being.

Throughout the city, the pageantry continued into the gray hours of morning when the stars faded in the approaching violet light. Far off in the distance, well beyond the waters of the Mississippi River, across the steel-colored Atlantic, a ship had set sail. On board, hundreds of mosquito eggs lay ready to hatch.

CHAPTER 2

Bright Canary Yellow

The *Emily B. Souder* steamed her way out of Havana headed toward New Orleans. She was what was known as a screw steamer with a cylinder engine and canvas sails taut against three oak masts. Built of hardwood and iron fastenings in 1864, she regularly sailed from New York to the Caribbean and up to New Orleans. The *Souder* was docked in the Havana harbor in the spring of 1878.

Havana was alive with ship traffic. Merchant ships from Boston, New York, Charleston, New Orleans and Brazil rocked in the slice of sea between El Morro Castillo and the seawall of Havana. A few of those ships had even arrived from West Africa. Though transatlantic slave traffic had finally been outlawed, ships carrying ivory, copper, palm oil and salt continued to make the journey. The *Emily B. Souder* was picking up a supply of sugar from a Havana wharf, but that was not the only cargo she left Havana with.

* * *

It was spring, and in West Africa, the wet season was under way. As the rivers and coastal areas were inundated with falling rain, mosquitoes proliferated. The mosquito found a perfect environment on board the oceangoing steamers: shelter, fresh rainwater, rotting fruit and an ample supply of warm bodies. The males, focused mainly on food and sex, fed off of the fruit and sought out their female shipmates. The impregnated females fed on the blood of human passengers in order to lay eggs. This transatlantic romance could repeat itself as many as three times during the journey.

In the water-dappled hollows of empty casks, the striped female mosquito deposited her eggs. Each time rain filled up the casks during the six-week sail across the Atlantic, the eggs hatched. When they did so, a virus born in the wilds of West African forests coursed through the new generation of *Aedes aegypti* mosquitoes.

Aedes aegypti, the striped house mosquito, looks like any other mosquito to the naked eye. Its elegant, gossamer wings flicker above a black body bright with white scratch marks. A silver lyre mark decorates its back. Its long, wiry legs are crooked high above, giving it the chilling appearance of impending attack. It is, however, heartier than many of its relatives. The striped house mosquito thrives indoors, feeding at any time of the day, and the females, who bite, outnumber the males five to one. The mosquito also has a peculiar adaptability to travel, prospering among the habitations of man, whether dwelling or boat. Some have been known to live for days in the damp clothes enclosed in a trunk. *Aedes aegypti* were native to Africa but established residence in the western hemisphere after centuries of ship trade provided the insect stowaway with repeated opportunity to colonize in the New World. The mosquito flourished.

When African ships dropped anchor in the Havana harbor, the new generation of female mosquitoes hunted warm-blooded mammals. The bright colors, movements, accents and sounds of the bustling harbor were lost on these insect immigrants; they were attracted to the ephemeral scent of exhaled carbon dioxide and lactic acid mingling in the humid air. Some moved unnoticed onto the shores of Cuba, where blood meals were in large supply. Others settled onto the decks of neighboring steamboats like the *Souder* where shipments of tobacco and sugar would soon depart for New Orleans. As the female mosquito departed her ship, she was attracted to the flailing arms and swatting that usually precipitate a swarm of mosquitoes. Her vision sensors locked onto the frenzied movement and its resulting heat. She landed on the flesh of an arm, easing in her proboscis, and injected a chemical that would prevent the human blood from clotting too quickly. As she did so, a sphere-shaped virus slipped into the bloodstream like oil entering water, and yellow fever replicated in the lifeline of an unsuspecting donor. The loaded mosquito simply moved onto the next warm body, where once again, she would exchange fever for blood. The infected person harbored the virus, unknowingly for a few days, while local *Aedes aegypti* mosquitoes then fed on the carrier human. This blood meal would pass the yellow fever virus into the gut and bloodstream of new mosquitoes, and the cycle, about one to two weeks long, would be complete. Within weeks, this rotation of the virus from mosquito to human to mosquito would create an insect population of virulent mosquitoes that would feed on the human population through all of the summer and fall. And so another yellow fever epidemic made its start in Cuba.

Though it was almost a yearly occurrence on the busy Caribbean island, this year's epidemic would prove to be unusually virulent, as though after two centuries it had finally perfected its genius for killing.

Yellow fever had been endemic in Cuba since the mid-1600s when the slave trade had established a sturdy colony of *Aedes aegypti* mosquitoes and a steady influx of the virus on cargo and slave ships. Small, sporadic cases would surface yearly. Exposure to these mild infections, especially during childhood, produced a type of immunity among the locals. Throughout the Caribbean and American South, it was almost a rite of passage to become "acclimated" to the fevers, and yellow fever quickly earned its reputation as a "stranger's disease" for its ability to hone in on new blood. In 1878, however, it was as though a new virus entered the circulation; the death toll mounted and those previously thought immune succumbed.

On May 19, 1878, the *Souder* set her course for Key West, where she would dock two days for supplies. This also provided the crew with some time for entertainment, mostly in the form of rum. Four days later, hungover and tired, the crew arrived at the Mississippi quarantine station outside of New Orleans and awaited inspection.

For three months the quarantine officers had been fielding cases of yellow fever coming from the Caribbean, in addition to four infected steamships from Rio de Janeiro. Two years before, in 1876, pressure from the New Orleans Chamber of Commerce, as well as a number of prominent physicians, had persuaded the Louisiana Legislature to weaken its laws on quarantine. Once required to spend ten days in detention to assure no cases of fever, vessels were now at the personal discretion of the board. To make matters worse, New Orleans officials were threatened with lawsuits if they detained a ship carrying perishable fruits. Many of those ships in 1878 came from Cuba, where the Ten Years' War for independence was coming to an end and an epidemic of yellow fever had been raging since March. Refugees landed in New Orleans by the hundreds.

The very day that the *Souder* arrived, another ship coming from Havana had declared five cases of fever on board. It was promptly taken into custody and would be held for nearly two weeks while it was thoroughly worked over with sulfur and carbolic acid. The harbor was filling with vessels bobbing in the water, the Yellow Jack flying high over their decks.

The captain of the *Souder* wasn't going to take any chances for delay; the last thing he needed was to be detained for a week or two at a quarantine outpost. A few of his men looked to be suffering from more than the average hangover, even complaining of fever. The captain met the quarantine physician on the gangway and offered up one feverish crewmember. The physician examined him, diagnosed it as malarial fever and sent him to the quarantine infirmary before examining the remaining crew. He recorded no other cases of fever. The physician did take notice of one other sailor looking sickly though; it was the ship's purser John Clark. The men blamed his condition on a rum hangover and "neuralgia," but Clark would later boast he "had beaten the quarantine officer." The *Emily B. Souder* was detained for only a few hours before she was given a clean bill of health and steamed her way into New Orleans, mooring in a berth off Calliope Street.

That night, John Clark's health worsened. Feeling feverish and agitated with an intolerable headache, he took a room at a nearby boardinghouse on Claiborne where a mulatto nurse looked after him. As his temperature climbed, his pulse slowed, known as Faget's sign. He felt intense heat all over his skin, but was unable to perspire. The fever attacked his organs, and his kidneys and intestines stopped functioning; high concentrations of uric acid collected in his kidneys, while he writhed from abdominal cramping.

His entire body ached from dehydration, and he suffered from severe hypoglycemia.

By Friday afternoon, Clark suddenly began to feel better and asked for food. A doctor who later described the scene wrote that the "fancy of food" always proved fatal. With the approach of evening, Clark again grew restless and his condition spiraled; the fever returned. He awoke twice in the night convulsing before finally slipping into delirium, his eyes glassy and empty. With the third convulsion, he died. It was shortly after 2:00 a.m. His dying liver had released a surge of bile, tinting the whites of his eyes and his skin saffron yellow. Upon later investigation, Clark was found to have been given treatments for yellow fever, though his death was officially recorded as malarial fever by the Board of Health.

Clark's body was removed, and he was quietly buried by 10:00 a.m. with no funeral and no public announcement of his death. Nearby streets were disinfected with carbolic acid, the smell lingering in the air well into the evening.

On the same day that Clark was buried, the *Souder*'s engineer, Thomas Elliott, fell feverish in a boarding room on Front and Girod streets; he died five days later in a nearby hotel. Rumor of another dead crewmember from the *Souder* caught the attention of two city physicians, and the body was sent to the dead house where an autopsy could be performed. The physicians described Elliott's body as "bright canary color" and his stomach filled with dark, blood-like matter. Dr. Samuel Choppin, president of the Board of Health of Louisiana, visited the dead house to examine Elliott's body. Choppin would later write: "These are all the usual appearances observed in the examination of a person dead of yellow fever, and we had no doubt that the man had been the subject of this disease."

It was the end of May, and yellow jack had gained its foothold in New Orleans. To a city on the bluffs, 400 miles to the north, this would have been urgent news. Since the two cities were first linked by railroads and steamboats in the 1850s, disease had been making its way into Memphis in fits and starts. By the mid-1850s the first railroad cars rolled between Memphis and New Orleans, and 200 yellow fever deaths soon followed. As Robert Desowitz, a professor of tropical medicine, wrote, "The railroads were viewed not as a channel of commerce but as a channel of contagion." Memphis experienced another epidemic in 1867 and a terrible one in 1873. At first, Memphians denied that the dreaded fever could have made its way into their promising city; the city was considered too high above sea level and too far inland for this plague of the port towns. To an extent, they were right. But once the railroads and steamships connected Memphis to the Gulf, the *Aedes aegypti* mosquito found its opportunity to move north. The insects settled near the river, finding refuge in the stagnant pools of water, private cisterns and the waxing and waning of the Mississippi riverbeds. The area at the mouth of the Wolf River was so hospitable to mosquitoes, one Memphian recalled, that a man could thrust his arm into the swarms of mosquitoes and leave a momentary vacuum when he withdrew it. Within ten years, there were enough striped mosquitoes in place to infect thousands of people in Memphis.

On the very day the *Souder* passed quarantine, a member of the Tennessee Board of Health wrote to the Louisiana health board in New Orleans asking for news of any yellow fever cases. Dr. Samuel Choppin, the doctor who examined the *Souder*'s dead crewmembers, assured the Memphians that they would receive regular reports and that nothing would be concealed from them. They did receive the reports; no mention of yellow fever cases ever appeared. Not until two months later, July 26, would Memphians read about an epidemic in New Orleans in the national papers. By then, it was too late. For all of June and July, New Orleans would routinely pass ships, including eighteen from Havana and fifteen from South American ports. In that time, the *Emily B. Souder* would make three more trips delivering sugar to New Orleans, escaping authorities and landing feverish crewmembers again in early July, sparking another wave of yellow fever cases. As the *Souder* unloaded cargo, other boats waited in nearby berths to sail north. One such towboat, the *John D. Porter,* left New Orleans on July 18 and entered the massive transportation waterway of the Mississippi River, where it would make its way north toward Memphis.

In the summer of 1878, the Mississippi River, the great artery of North America, would carry death in its bloodstream, spreading the worst yellow fever epidemic in American history.

CHAPTER 3

The Doctors

The sweet scents of honeysuckle, overripened fruit, and fresh-cut hay distracted from the fecund smell of the living body of water coursing beneath the bluffs of Memphis. Crape myrtle bloomed around the marble fountain in Court Square. Geraniums, roses and lilies bejeweled the park, and magnolia limbs bowed under the weight of the dark, glossy leaves. Throughout the park, the calls of the bootblacks offering shoeshines could be heard, and cicadas hummed from the treetops.

Early in the morning, before the haze of summer humidity settled over the town, the voice of the milkman resounded: "Wide awake," "All alive, now!" The ice wagon rattled up Second Street, while the produce stands opened in front of the public offices and dry-goods stores on Main. Dust, mud and manure along the bustling thoroughfares necessitated that even the most well-bred lady wear rubber Wellingtons downtown. Nicholson wood-plank sidewalks, which had been rotting for ten years, were piled high with crates where newsboys sat shouting headlines.

When the summer heat arrived early, as it did in 1878, it was known as yellow fever weather. Memphians passed one another downtown with nervous glances, their eyes averted beneath cotton bonnets and straw hats. Few uttered the words "yellow fever," but the fear of it suffused the still air nonetheless. In addition to the early arrival of summer, the high temperatures and little rainfall had produced a drought, draining the Mississippi River down to the hardpan. Green lawns lush from the surplus rain in months past turned as brown as prairie grass, and water wells dried up. The scant rain had left no natural way to wash sewage from the dusty Memphis streets, and bloated, dead animals rotted in the shallow pools of refuse; the city had no money to remove them. Finally, rumors of a yellow fever epidemic in New Orleans spread up the Mississippi Valley, and in Memphis, the talk of fever consumed conversation like brushfire.

Yellow fever answered to many different names, including yellow jack, coup de barre (blow of the rod), vomito negro (black vomit), the saffron scourge and the American plague. Regardless of what it was called, yellow fever, elusive and terrifying, remained a mystery to medical minds during the nineteenth century. Disease in general was an enigma, medicine still in its infancy. By the 1870s many doctors received no formal medical education, instead receiving their knowledge through a type of medical apprenticeship with local physicians or during the Civil War. The medical programs that did exist at places like Harvard and Yale had no admissions program, performed few autopsies, did not even own microscopes. While Europe excelled in its medical research, American institutions placed almost no value on evidence-based medicine. To them, the body was a balanced system, and their work focused on treating the occasional imbalance with lancet bleeding, leeches, castor oil and even arsenic at times.

To make matters worse, a thin line existed between medicine and religion. Newspapers smacked of advertisements for medications like Tutt's pills, which could cure everything from wind colic to low spirits, and were "Recommended by physicians. Indorsed by Clergy." The physical body was the realm of the Almighty, so knowledge of its inner workings through autopsy or examination was like trespassing. From the pulpits people were often told that epidemics, *plagues*, resulted from wicked ways and intoxication. And, in reality, there might have been some common sense to it considering venereal disease was one of the largest health concerns of the day; a number of doctors even specialized in these private or *secret diseases*. Very little connection had been made between filth and disease, so amid this near witch-doctor approach to medicine, cities remained baffled by cholera epidemics as they drank from the same sources in which they dumped sewage. The germ theory would soon change all of that, but in Memphis in 1878, sanitation, like immoral behavior, was still just one in a number of guesses as to what caused and spread disease.

Doctors relied on two prevailing theories about yellow fever: One camp believed it was mysteriously spread by filthy conditions much like cholera and dysentery. Terms like fomites, effuvia, and noxious gases peppered medical literature in an attempt to explain what substance—whether animal matter, fungal or gaseous—spread the disease. The other side held that the fever was imported each summer into the city by railroads and river traffic. Not sure which would prove to be more effective, Memphis health officials decided to tackle both problems.

In the summer of 1878, the Board of Health secured $8,000 in city funds to clean up the foul city, investigating cisterns and outhouses. Wayward goats and hogs were impounded. Regulations for the disposal of animal carcasses were strictly enforced. Regardless of their efforts, over 70 percent of the city's dwellings were wooden and prone to rotting. When it rained, basements flooded, holding several inches of water, while the walls wept polluted mud. The Nicholson paving continued to decay, and the city's water, though routed through the pumping station, came from the Wolf River. Even milk, under no inspection, was diluted with river water, and it was reported that one person found a minnow in his milk jug. The biggest problem was that of raw sewage and privies. Like all other densely populated cities, there was no effective way to remove waste. Citizens in downtown Memphis carted and dumped their privies into the river, and over half of the city's privies were located fifty feet or less from drinking wells. One authority would call privies the "most annoying problem connected with urbanization."

Nonetheless, as yellow fever season loomed, one newspaper column read, "Memphis is about the healthiest city on the continent at present," and another bragged, "We need not fear in Memphis. We were never in as good a condition from a sanitary point of view . . . Nothing in our atmosphere invites that dread disease." In the greater scheme of things, from political neglect to public apathy, the city of Memphis was due a tragedy, though no one seemed ready to acknowledge it.

The war on disease was a two-front campaign. Despite all their efforts, the Board of Health had failed in their campaign to clean the filthy city. They turned now to the issue of quarantine.

Quarantine had been a regular measure for preventing disease as early as the biblical lepers, but it was most widely used beginning with the bubonic plague in fourteenth-century Venice. Ships were forced to spend weeks anchored outside of a city until the crew showed no signs of disease among them. Vessels and their cargo were initially intended to spend thirty days—*trentina*—in the harbor, but that later changed to forty days—*quarantina*. Quarantines continued to rule maritime travel well into the nineteenth century, finding even greater cause with the trafficking of human cargo. European ships often spent weeks moored off the coast of Africa before crossing the Atlantic. Fever was so prevalent among the crews that one island, São Tomé, was known as the Dutchman's graveyard. And the tale of the Flying Dutchman is thought to be the story of a yellow-fever-infected ship repeatedly denied port until all on board perished of the fever, and the ship was forced to sail endlessly, manned by a ghost crew, delivering detriment to other seafaring vessels.

Quarantines were not limited to maritime trade, however. In America, trains and paddleboats could also be quarantined to prevent smallpox, plague and fevers. Even returning soldiers were quarantined in camps. And once a city was known to harbor infectious disease, the town itself could become quarantined from the rest of the nation. Fleeing citizens were denied entry into healthy cities. Even Alexander Hamilton came down with yellow fever and was refused entry into New York City. He found refuge in the country until he recovered.

By the mid-nineteenth century, merchants and businessmen were becoming frustrated by the practice of quarantine, which proved more effective in preventing the delivery of goods than disease. Port towns in particular had become lax in their quarantine measures, yielding to commerce. On April 29, 1878, Congress had finally passed the Quarantine Act granting the Marine Hospital Service quarantine rights along port cities. If local governments could not be depended upon to enforce quarantine, the military could. The law, however, was a weak one and would take several months to go into effect. For the Mississippi Valley and beyond, the delay would prove to be disastrous.

In what would become known as the "war of the doctors," the leading medical minds in Memphis debated quarantine. Dr. Robert Wood Mitchell was forty-seven years old, his white goatee brushing against a stiff, detachable collar and black bowtie. Mitchell had served as a division surgeon in the Confederate Army of Tennessee, and he wore the terror of that war in lines that feathered from his eyes and brow. Mitchell was described as "marvelously tender and sympathetic," as well as "born to command . . . so great is his power and influence over men." But it was an unassuming influence. If men followed him, it was because of his unclouded allegiance to serving—whether in battle, medicine or tragedy—rather than a charismatic personality. After the war, Mitchell returned to Memphis where he married an Irish woman and set up his practice Mitchell and Maury. He earned a reputation for excellence in his field, and in March of 1878, one week after the Mardi Gras festivities came to a close, Mitchell was appointed president of the Memphis Board of Health.

The board consisted of three doctors, the chief of police and Mayor John R. Flippin. In early July, the five men met to discuss how to deal with the 1878 epidemic season. As they opened windows to give way to the breeze and paced on plank-board floors, the hot season was upon them: Heat lightning cracked the nighttime sky, sermons were shorter on Sundays, and the picnic season had been declared officially over. Ice-cold buttermilk had become the fashionable drink. Watermelon, cantaloupe and peaches softened and seeped from crates at the fruit stands. Corn was on the silk, and advertisements for refrigerators, ice chests and coffins filled the pages of the newspapers. It was also noted in the paper that "Mosquitoes are increasing in numbers, and are becoming more vindictive and ferocious, if it were possible to do so."

Mitchell called his meeting to order. He was a quiet man, a listener who rarely spoke unless he had something important to add to the argument. He had already appeared before the local council in June to formally request additional money for sanitation and quarantine as the epidemic season approached. His request was denied. A staunch believer in annual quarantine during the summer months, Mitchell then appealed to his own board for quarantine. The majority voted yes; but the two other physicians on the board voted no.

Dr. John H. Erskine listened to Mitchell's pleas. Erskine was a commanding presence, standing at an unusual six feet tall. His close-set, pale eyes and fair skin glowed against black-blue waves of hair and a long, dark goatee. Erskine's height coupled with his striking appearance gave him an air of leadership; he was used to people not only taking notice of him but listening to him. A highly educated and ambitious surgeon, he excelled during the war and later as the health officer of the Memphis Board of Health during the 1873 yellow fever epidemic. During that epidemic, Erskine had even noticed an unusual occurrence: The prison where he often worked reported only two cases of the fever. The fifteen-foot-high prison wall, Erskine noticed, had somehow barred the fever from entering.

Still, as Erskine listened to Mitchell's pertinacious pleas, he did not believe the rumors from New Orleans warranted such drastic measures. Quarantine, after all, would create panic, stifling river traffic and delaying cotton shipments. It was not even a proven method of protection against the scourge. Erskine spearheaded a petition, signed by several prominent physicians and published in the newspaper, overturning the vote for quarantine.

It brought the battle to the public. Angry letters to the Memphis *Appeal* asked, "Is it not better to expend a few thousand as a safeguard than to lose millions by the disastrous effects of yellow fever, besides the thousands of valued lives that will have passed away?" The paper followed with an editorial: "Should an epidemic reach Memphis this year those who opposed the establishment of a quarantine will be held responsible."

It is impossible to know what went through Mitchell's mind at that moment, frustrated by his own board and fed up with city officials. On July 11, Mitchell resigned from the Board of Health, and in spite of a 400-person petition in support of him, Mitchell would not change his mind. To replace him, the mayor instated Dr. Dudley Saunders, who saw no need to quarantine the city and choke river traffic.

The Memphis *Appeal* published a letter from Mitchell in which he explained his resignation to the people of Memphis: "I may add that at present I see no cause of danger from yellow fever; but we saw none in 1873 . . . It is my earnest and honest conviction that should we ever have yellow fever again, it will be our own fault in not taking the known necessary precautions against it."

In the coming weeks, both Mitchell and Erskine would think back to their initial actions and that fateful vote—one would survive and one would perish in the consuming epidemic. The "war of the doctors" had been a futile skirmish, waged too late to be effective. Unknown to any of them, the fever had already taken its first victim in Memphis.

The *Appeal* also published an editorial in favor of Saunders: "The public will hail his appointment with satisfaction and feel assured that with him and Dr. John Erskine . . . all will be done that can be done to keep our city free from epidemic disease." On that very day, national news of yellow fever in New Orleans finally prompted the Memphis Board of Health to reconsider their long-argued decision and establish quarantine. Mayor Flippin assigned a physician, one who had signed the petition against quarantine, to the quarantine post on President's Island in the Mississippi River. It was July 27.

Police with shotguns stood along the train tracks of the Mississippi and Tennessee railroads at Whitehaven Station, eight miles outside of the city. The Charleston railroad at Germantown, a good twelve miles away, was quarantined. Cotton, sugar and coffee shipments were held, and all river ports closed with posted shotguns, citizens, believing as they did, that firearms and willpower might be enough to barricade disease.

The board resolved to meet every Monday night at 8:00 during the yellow fever *excitement*, and Dr. Erskine would purchase barrels of Calvert's No. 5 carbolic acid to use as disinfectant. Nonetheless, the quarantine did little to ease the minds of the people. Restless in the quiet before the storm, Memphians seemed to feel the epidemic long before any confirmed cases of disease surfaced. Bank accounts were cashed out, letters were written to relatives in other cities announcing plans to visit, businesses kept shorter hours. The heat always made it difficult to sleep, but now worry added to the collective insomnia. Even the animals furthered the subtle sense of hysteria. Throughout the town, people noted that pets and farm animals had been running away; the birds had stopped chirping. The only life-forms unfazed were the unusually high number of striped house mosquitoes.

July had also been a month of exciting, if not strange, occurrences. First a well-known citizen had been crushed beneath a streetcar amid all the celebration and festivities on July 4; another had been struck by lightning just a week later. Streetlamps ex-

ploded and caught fire, while mules fell dead of heat stroke in the
fields. Citizens gathered downtown to listen to a technological
breakthrough, the Edison speaking phonograph. Then, it was reported that a five-and-a-half-foot rattlesnake had been killed in
Memphis. The strangest of all, however, was the approaching solar eclipse. In a fit of brevity, there had even been two new cocktails created that July—one was called the Quarantine, the other
the Eclipse. The Quarantine, it was said, would isolate and insu-
late from all other drinks; the Eclipse would shut out what your
neighbors are taking in theirs.

For those who followed the stars, the solar eclipse would occur during the constellation Ophiuchus, the serpent carrier, and
the little-known thirteenth zodiac. Ophiuchus is the only constellation based on a real man, a doctor who lived in ancient Memphis, Egypt, and a solar eclipse under his sign was said to bring
disease and struggles between life and death. Another solar
eclipse happened under his watch in Philadelphia in 1793 just before an outbreak of yellow fever. That epidemic, one historian
wrote, "ushered in the longest and deadliest string of yellow fever
years yet known in North America."

On the afternoon of July 29, Memphians braved the ninety-
degree heat to stand in the streets and along the bluffs at 4:28 in
the afternoon. They held opera glasses and smoked glass heavenward and watched the moon cross the sun. An onyx sphere crept
partway across the bright light, and to the spectators on the bluffs
of Memphis, there seemed little question as to which was
stronger as the dark eclipsed the light. It was like dusk had suddenly descended upon the city.

When yellow fever first arrived on the streets of Memphis, it did
so silently. On July 21, a man arrived off of a riverboat to visit his

wife who worked as a cook at 279 Second Street. The Victorian home was the downtown residence of Attorney General G.P.M. Turner. The man fell feverish, but soon recovered. Ten days later, Turner's two children burned with fever. One survived, the other died.

July 25, Willie Darby, an employee of Farrell Oysterdealers who lived at 277 Second Street was taken with a fever and soon recovered. Though no doctor confirmed it, Darby's was the second known case of yellow fever in July of 1878.

Other unconfirmed cases along Second Street surfaced, and Memphis police learned that travelers had evaded quarantine in South Memphis, a suburb that had recently been incorporated, slipping by night into the unguarded yard of the gas works.

Then, the plague-stricken steamboat *John D. Porter* passed up-river with her tow at the end of July. Crowds of Memphians, who had heard that the *John Porter* carried yellow fever from New Orleans, gathered along the river to watch it approach their city. Dr. John Erskine was called in under the laws of quarantine. He rode a tugboat across the river and boarded the ship. The captain, denying any cases of yellow fever, reported that four men had died and one crewmember was sickly from sunstroke. It is unknown whether or not Erskine saw any of the dead crewmembers, nor if he saw any obvious signs of yellow fever on board. What is known is that the suffocating confines of the cabin made it easier to blame the deaths on overheating. The *John Porter* was asked to by-pass Memphis, and the towboat would continue up the Mississippi River, spreading yellow fever all the way to Ohio, where its crew finally abandoned the cursed boat.

On August 1, the steamer *Golden Crown* landed three feverish ladies on the banks of Memphis. The same week, the newspaper reported, "The city is quiet and yellow fever rumors appear to have abated ... This should effectively dispose of the tale circulated by sensationalists about the presence of yellow fever here."

In hindsight, it seems negligent that the board and the press would be so quick to dismiss the possibility of another epidemic. Epidemics hit in 1867 and again in 1873, the latter taking 2,000 lives in Memphis. But it also proves true of the time period, one in which men began to see themselves as governing over nature itself. Hubris abounded. And as was the case with most American cities, civic and business leaders alike looked more at commerce and industry, new roads and railroad tracks, than at sanitation or poverty-ridden districts. A *New York Times* columnist writing of the devastating epidemic in the South editorialized, "It would have been only necessary for the deadly germs to get abroad in one of our filthy tenement house districts to spread terror and dismay and defy all human efforts to exterminate it." In a struggling city like Memphis that depended heavily on river traffic, another epidemic, or even word of one would be costly. The board unanimously promised, however, that should symptoms of yellow fever arise, they would make no attempt at concealment.

On August 1, a man named William Warren, a deckhand from the *Golden Crown,* landed in Memphis and visited an Italian snack shop in the Pinch District. Named for the "pinch-gut" appearance of its hungry Irish immigrants, the Pinch was the closest neighborhood to the river, wedged between the Wolf River and the Gayoso Bayou. Here, the river traffic congregated, and it was not uncommon to hear drunken laughter and the roll of bone dice. Gambling, prostitution and drunkenness flourished. The next morning, William Warren grew feverish. Dr. Erskine was notified once again and admitted the man to a quarantine hospital on President's Island where he died three days later of fever. Though yellow fever proved hard to distinguish from other tropical diseases at its onset, it was unmistakable at the time of death. The deep yellow skin appeared like tarnished brass, marred by violet-colored spots. Still, Warren's case never went on record.

The first case to go on record for the public was not for twelve more days when Mrs. Kate Bionda, owner of the Italian snack house Warren had visited, died of the fever on August 13. Hers was officially reported by the Board of Health, on August 14, as the first case of yellow fever in the city. Bionda's shop, a clapboard shack reeking of fish and catering to the river traffic in the Pinch, was fenced in, blockaded. All adjacent buildings were disinfected. The body of Mrs. Bionda was burned within five hours of death. On that day, the *Appeal* published a warning: "The sad case of Mrs. Bionda, who left two little children and a grief-stricken husband, does not prove necessarily that others will follow. There is no need of a panic or stampede." Only two cases of yellow fever went on record that day; at least twenty-two cases existed.

YELLOW FEVER IN MEMPHIS flashed across telegraph wires nationwide. It would chill residents from New York City to Philadelphia to Charleston and every town, small or large that fed from the Mississippi River.

The same trains and steamboats that brought thousands into Memphis for Mardi Gras that spring now carried away over 25,000 Memphians, more than half of the population, in a span of five days. In mass exodus, the rich fled by train, carriage, or boat, leaving dinner tables still set with silver and doors wide open. Traffic clogged the roads, and dust sprayed in the wake of carriage wheels. At the depot, ticket sales in one line alone exceeded $35,000. Platforms were piled high with trunks, suitcases and furniture. In an age of honor and Victorian manners, people trampled over one another. As one man commented, "For the sake of humanity, men became inhuman."

The city collapsed, hemorrhaging its population, its income, its viability. Trains pulled away, leaving people weeping beside the tracks, their last chance at escape gone as the final train cars rolled to a start. A morbid calmness fell over Memphis, so still and quiet as to be serene if one didn't know it was simply the pallor of death. In July of that year, the city boasted a population of 47,000. By September, 19,000 remained and 17,000 of them had yellow fever.

Once free of the city, Memphians did not fare much better. Nearby farmers locked their gates and doors, with shotguns ready. Public roads were wrecked and bridges burned to prevent travel. Many cities and towns refused admittance in fear of the dreaded fever. People were crowded into train cars with no food or water, the smell of sulfur still fresh on their shoes. A man jumped off a train to fill a bucket with fresh water at one stop. He sold the precious commodity to passengers for one dollar a cup at a time when a dollar was considered a decent day's wage.

One train sent a message to the nearby town of Milan for food and water. The townsfolk set up tables of food and tubs of water along the river bottom, four miles outside of town limits. The people gathered on a hilltop a mile away and watched as the men, women and children released from the train cars ran toward the provisions. Then, guards with shotguns herded the Memphians back into the suffocating train cars. Those were the lucky ones. Left behind in the ruined city were the poor, the sick and the dead.

Immediately, a Citizen's Relief Committee, headed by

Charles G. Fisher, convened in the Memphis opera house. Since most wealthy city officials had fled, ordinary citizens took up the call. Their first priority: to get the poor out of the city and organized in refugee camps. Already, families sought out chicken shacks and abandoned slave quarters throughout the countryside. The committee turned every stall and booth of the county fairgrounds into shelters. It was said that the flutter of canvas could be seen on every grove in Memphis.

The Howard Association, formed specifically for yellow fever epidemics in New Orleans and Memphis, organized nurses and doctors. They asked Dr. Robert Mitchell, newly resigned from the Board of Health, to be medical director. He accepted the position in what must have been a state of despair. Mitchell had known this would happen; he had seen it coming and could do nothing to stop it. His fate, it seemed, would be fixed to this fever.

Amid shimmering light and supple breezes on August 23, the Board of Health finally declared a yellow fever epidemic in Memphis. It was two months after Mitchell's ill-fated battle for quarantine; one month after the fever's first victim.

CHAPTER 4

A City of Corpses

Dust from the lime blew in the air, bone colored and sifted fine as flour. There was no traffic to stir it up. No horses to kick dirt beneath their hooves. It rose and fell of its own energy, occasionally accompanied by the pitch of a mosquito's wings.

Only six months after the lavish Mardi Gras celebration, Memphis was a city of corpses. Streets, white with disinfectant, were deserted. Once lined with cotton bales and parade bleachers, Main Street now held piles of coffins, stacked one on top of another, so that walking the thoroughfare felt like entering a tomb. Instead of pageant masks, the occasional pedestrian would hurry by with a sponge tied across his nose to cover the smell. Where 3,000 horsemen had paraded, only death wagons now clattered by in pairs: One wagon of empty coffins, the other of full ones. The smell of cologne and rosewater, sprinkled on the bedclothes of the dying, seeped from doorways disguising the peculiar, pungent odor of illness. The sweet scent of blossoms had been replaced by the saccharine stench of death.

Wilkerson Drugs had closed, as had most other apothecaries, their colorful show globes void of light. Prescriptions could not be filled. McLaughlin's grocery and Barnaby's shop were boarded shut. Vegetable carts had long since closed, and the milk wagon had ceased its rounds weeks ago. Banks opened for only one hour a day. Memphis, having been quarantined from the rest of the country, became a colony left to burn. One journalist wrote: "A stranger in Memphis might believe he was in hell."

City officials wired President Rutherford B. Hayes for help; little was given. Hayes wrote in a personal letter on August 19, "I suspect the Memphis sorrow (yellow fever epidemic) is greatly exaggerated by the panic-stricken people. We do all we can for their relief." On September 2, Mayor Flippin again telegraphed the president for assistance, but it was the last of such correspondence. Four days later, the mayor was down with the fever.

Even the lime could not cover the smell of death as Constance stepped off the train platform on August 20, 1878. The wind carried the odor for three miles outside of the city. Sister Constance and Sister Thecla returned from a vacation on the Hudson as soon as they heard the news of the fever; the sisters were the only ones traveling into Memphis.

As they made their way through the town, signs of plague were everywhere. Across the street from the marble fountain of Court Square stood a white, clapboard building flanked by two staircases. It was the headquarters for the Memphis Board of Health. In front of it, wagons filled with disinfectant held shovels protruding out of the flatbed like broken limbs. On a trip through the city the shovels would empty the chalky chemical as downy as falling snow; on the return trip, the shovels picked up badly decomposed bodies.

The carriage pulled away from the downtown train station, up Poplar Street past the empty courthouse on Main. It moved slowly through the streets, navigating the huge sinkholes and corroded paving. The smell of the Gayoso Bayou and all its decay was heavy in the air. A hot breeze lifted the treetops and, already, the leaves began to burn at the edges. In spite of temperatures that hovered around 100 degrees, residents had been advised to keep fires burning within their homes to cleanse the air, and windows were boarded shut against the pestilence.

As the sisters entered the infected district, yellow pieces of cardboard marked the doorways of the ill. On many porch fronts, black replaced the yellow cardboard with white chalk scrawled across it—*Coffin Needed*—and the dimensions for a man, woman or child.

It was a pitiful parting in a time of extravagant mourning. Under normal circumstances, the dying family member would have had the opportunity to say good-bye to all loved ones as they gathered bedside to hear the last words. The family would then have drawn the blinds, covered mirrors in black crepe and stopped all of the clocks. Strands of the deceased's hair might be cut and woven into shapes like a cross to display in a glass case in the parlor. Even the children and babies would take part in the mourning, wearing a touch of black. The body would be packed in ice if it was summer and laid out in the parlor—a tradition that with time would dwindle, and the term *parlor* would be replaced by the *living* room. Finally, the women would stay behind in the home, while pallbearers in black gloves carried the coffin to its place of burial, where it would be draped with fresh flowers. Formal announcements of death would be mailed. And the widow would forgo any gold or silver jewelry, wearing a dark veil during the following year and black garments for the next two and a half years.

During the epidemic, however, families prepared their own for burial, cleaning the bodies when there was time, placing the corpse in a pine box with a mixture of tar and acid before bolting the lid closed. They would listen. At some point during the day, in the suffocating silence, a team of six horses pulling a wagon would come up the block and announce, "Bring out your dead!"

The infected district started with the river. From there it spread through the lowland just underneath the bluffs and Front Street known as Happy Hollow. At Happy Hollow the Wolf River joined the muddy Mississippi, creating a rich stew of bog land and river brush. Happy Hollow had been the primary dump for downtown citizens who still relied on the bucket-and-cart system for emptying their privies. The mixture of refuse, rainwater and mud created a landfill where poor immigrants could build makeshift homes out of boat scraps and sheet metal perched on stilts above the fetid mud and froth. The accommodations were both rent free and tax free.

For two decades, railcars and steamboats had transported more than goods to Memphis, they delivered immigrants looking for work. Around New Orleans, yellow fever, the "stranger's disease," regularly fed off these newly arrived, nonimmune immigrants. In the 1870s, a large influx of immigrants moved into Memphis and settled around the river in Happy Hollow and the Pinch District. To a virus preying upon populations of nonimmunes, they provided ample supply.

From Happy Hollow and the river, the infected district spread across Front Street into the Pinch, then Second and Third streets.

It stretched south into Exchange, Poplar, Washington and Adams streets. Deep within these neighborhoods stood the Memphis Courthouse, Calvary Church, Grace Church, a synagogue, City Hall and an elaborately expensive prison. Of all the dwellings, the prison would report the fewest cases of yellow fever.

As Constance and Thecla made their way through the infected district, they crossed the Gayoso Bayou, and at last, reached Alabama Street where St. Mary's Cathedral stood, a wooden, Gothic church with a large rose window over the entrance. Petals of purple, blue and gold shone light into the dark wood interior where a tall, arc-like nave gave the feel of a ship turned inside out. The church, now a bishop's cathedral, had been built in the 1850s as a branch of Calvary Episcopal Church and St. Lazarus-Grace Church. It had been constructed on the very edge of town where Poplar intersected Alabama Street and Orleans with instructions for a steeple that could be seen from Main Street. At a time when other churches charged fees for their pews, St. Mary's did not, hoping to be open to all people, those in the city and those in the country, the ones who could afford it and those who could not.

The conditions at St. Mary's were not much better than those of the town. Already home to a girl's school and church orphanage, the Citizen's Relief Committee then appealed to the sisters of St. Mary's to take care of the Canfield Asylum, a home for black children, as well. Hundreds of children orphaned by the epidemic took residence at Canfield, one of the last disease-free havens in the city—far from the Mississippi River unfurling the pestilence from its banks.

"What we've decided to do is have you sleep in the country, out of the infected atmosphere," explained one of the sisters from St. Mary's. "You can work in town during the day."

Constance and Thecla refused. "We cannot listen to such a

plan; it would never do; we are going to nurse day and night; we must be at our post."

Both sisters had survived the previous yellow fever epidemic in 1873. No sooner had they arrived in Memphis to open a boarding and day school for girls than the epidemic began. Women trained to be teachers found themselves to be nurses, cooks, caretakers. One sister at St. Mary's would later write: "That epidemic of 1873 seems now like a faint foreshadowing of the one through which we have just passed. The distressing scenes witnessed in the first were replaced by overwhelming sorrows in the second, while the pain and sadness of the one were intensified into most bitter suffering and anguish in the other."

When a scourge of this magnitude strikes, the minds of people, against all rational thought, look for a reason. Modern-day epidemic psychologists have described a total collapse of conventional order—fear pervades, the sick go uncared for, people are persecuted and moral controversies arise. Memphians became almost medieval in their divine conclusions. Protestants and Catholics fought over who deserted and who stayed to look after their flocks. It was suggested by some in the North that an *all-wise Providence* created the plague to bring a divided nation together. Clergy warned that New Orleans and Memphis suffered yellow fever because of their heathen Mardi Gras celebrations. The fever, it was reported, "was fatal to those whose energies had been exhausted by debauchery." It made it harder to explain the many who perished from selfless sacrifice. "The nuns died," as one newspaper column read, "in numbers sufficient to give rise to the belief that they were specially marked by the destroyer."

Constance served as sister superior at St. Mary's. Caroline Louise Darling, as she was named at birth, was accomplished for the time period: educated, talented, well mannered, a good leader to the band of women at St. Mary's. She was described as "a woman of exquisite grace, tenderness, and loveliness of character, very highly educated, and one who might have adorned the most brilliant social circle." Constance looked young for her thirty-two years with a round face and blue eyes beneath the heavy black habit, an iron cross around her neck.

When Constance arrived at St. Mary's that August day, she went immediately to meet with the dean, Reverend George C. Harris. The greatest task facing them was caring for the dozens of children orphaned daily, but their help was also needed in the streets. The virulence of this particular yellow fever epidemic was without question, but neglect of the ill proved to be almost as deadly. Nurses were scarce. Many patients, who might have recovered, died simply from starvation and dehydration.

Dean Harris outlined their immediate needs: to feed the hungry, to provide for the barest necessities of the sick, to minister to the dying, to bury the dead and to take care of the orphaned children.

Each day, the sisters alternated caring for the orphans at St. Mary's, delivering children to the Canfield Asylum and taking soup and medicine on house calls. They met in the mornings with Dean Harris to receive their orders, then set out into the infected district with linen squares soaked in disinfectant hidden beneath their clothes. Every evening they met for Communion and Vespers, and at Harris's request, they relaxed without talk of the fever.

Constance must have set out her first day with a sense of purpose and strength. Stepping into the heat and stench of Poplar Street, it would not take long to realize the overwhelming magnitude of this scourge. The sun, dust and black smoke of fever fires coiled around her in a dizzying haze. Where to begin? Which neighborhoods, which streets, which houses? There were too many to count. Victims fell dead in the parks, under fences or alone in their homes, only to be discovered when the August heat picked up the scent. "Some," the paper published, "were found in a state little better than a lot of bones in a puddle of green water." Children were found sick in the same bed as their deceased parents. One mother was found dead beside a starving infant still trying to breastfeed.

A man met Constance in the street with a telegram. It was surprising to receive such an official note. Only days before, the newspaper had published a notice from the telegraph company requesting people to pick up their own messages; all of the messengers had left the service of the agency.

The messenger handed Constance the telegram. *Father and mother are lying dead in the house, brother is dying, send me some help, no money,* signed *Sallie U.*

"Will you go to that poor girl?" he asked.

A number of nurses, doctors, ministers or nuns later wrote of the fear that accompanied them the first time they entered an infected home. They had nursed hundreds from the halls of sick wards, but it was something else all together to climb the steps of a porch and open a door with a yellow card swinging from a nail. The first thing to strike was the smell. It floated into the streets, a scent like rotting hay. The smell grew stronger and overpowering once the front door opened, where it mingled with soiled sheets, sweat and vomit. Inside, one never knew what to expect. Moans, cries, delirious screams, or worse, no sound at all. There was darkness, as windows were boarded shut, and there was the stagnant heat of imprisoned air. Then, as their eyes focused, they saw the bodies. At first it was hard to tell which ones were living and which were not. If deceased, one could never know how long they had been that way or in what condition they would be.

Constance arrived at a small but neat home. Serpentine watermelon vines grew wildly around the homes in the neighborhood, and abandoned cats and dogs howled for lost owners. A pretty young girl in mourning led her into the house. Dust floated, effulgent, in the shafts of afternoon light, and the air was heavy as steam. One corpse lay on the sofa, another one on the bed, their skin yellow and tongues black. A tall young man, nearly naked, was also in the bed, delirious, rocking back and forth. His eyes sank deep into his cheekbones ringed by bruised half moons. Outside the window, Constance heard a crowd gathering, presumably to loot the house once all were dead. Constance ran into the yard and shouted at them to leave, warned them of the plague. They scattered like insects in the sunlight.

The healthy were not permitted to touch the dead for fear of spreading the disease further, so Constance sent for an undertaker. But, it could take as long as two days to have the bodies removed. Mr. Walsh, the county undertaker, refused to pay extra wages to the *colored men* loading and unloading the bodies. Finally, he was arrested. From then on, the men were promised five dollars for an adult corpse, three dollars for a child. In the meantime, the Citizen's Relief Committee arranged burial patrols to locate bodies by report, smell or even the low flight of buzzards. At the hospitals, patients died so quickly that thirty new corpses might be piled in the dead house before the undertaker returned from the cemetery.

The grounds of Elmwood Cemetery were bloated with shallow graves, some only sixteen inches beneath the surface. Deep, muddy scars cut into the grounds where coffins had been laid side by side in long rows in the earth. And on more than one occasion, a knock was heard before the lid was screwed tight or the coffin lowered into the ground, and a patient, thought to be dead, would call out from inside.

Elmwood was two and a half miles outside of the city between a railway line and North Walker Avenue. A streetcar ran from the city to the cemetery every ten minutes where visitors with admission tickets could visit family plots. Weeping willows, seashell roads and flowers made the cemetery a peaceful place of recreation. Families purchased plots at Elmwood—an adult, white, first-class plot cost about fifteen dollars, while an adult, black, second-class plot cost around twelve dollars. Headstones could cost anywhere from two and a half to seventy dollars. Cemeteries had long ago moved away from church graveyards to larger land holdings outside of cities to prevent the spread of disease. Still, Elmwood strictly enforced its rule about internments—a body could only be moved during the months of December through March, considered the non-epidemic months, unless the body had been dead five years.

A young girl named Grace lived with her father, the superintendent, in a cottage at Elmwood. She tolled the bell each time a body was buried and kept the names in a large, red leather logbook. During the month of September, there was page after page of yellow fever victims. It was said that the bell at Elmwood tolled constantly that month.

Constance left the small house sick from the stench. The air, suffused with moisture, closed the odor of death around the town and its people. She went in search of more nurses and beef tea for the ill. As she did so, she noticed a spectacular sun, a blood orange setting over the Mississippi. How strange, she thought, that one could still find anything beautiful at all.

By dusk, plumes of black smoke climbed the night air as contaminated mattresses, blankets, furniture and clothing burned in the fever fires. Fire engines hosed the streets in attempts to wash them. A mixture—three pounds copperas, one pint of carbolic acid, one bucket of water—ran through the alleyways and cesspools for cleansing. When rain began to fall, blasts of gunpowder quaked throughout the city to clear the air of disease. The chief of police had ordered a curfew for the citizens and the midnight church bells silenced. In the stillness, tree limbs cast ominous shadows like awnings over the streets, and in the deepest hours of night, the city seemed to drone with the sound of dirges.

At daybreak, smoldering piles of charred bedding were silhouettes against the lawns and tidy houses of the dead. People climbed into their beds at night wondering if their own belongings would be burning on the lawn the next day. At one home, the sisters found two bodies still unburied, rotting. This time, Constance found a policeman to secure the undertaker. At another house, a mother lay dying, leaving four children and a baby starving. Constance drove through town looking for milk to feed the infant. Eventually, the horse lost all of its shoes. There was not one blacksmith left in the city.

Constance wrote to her mother superior. The post office was still in operation, though letters postmarked from Memphis arrived with five or six holes punched through them. A nail-studded paddle pierced the paper to "fumigate" it with a solution of sulfur. Constance's letter implored the mother superior for more help and described the day-to-day siege yellow fever took on the city: "One grows perfectly hardened to these things—carts, with eight or nine corpses in rough boxes, are ordinary sights. I saw a nurse stop one today and ask for a certain man's residence—the negro driver just pointed over his shoulder with his whip at the heap of coffins behind him and answered, 'I've got him here in his coffin.'"

It was one of the few letters she wrote; Constance worked constantly. On some days she kept a diary, but on others, she worked straight through several nights in a row. One of her sisters pleaded with her to rest and eat something.

"Sister, I am hungry all the time, no matter how much I eat: I am so very well. Do not worry for me." She sat down to a small plate of food, brushing at the flies and sweat bees hovering in the humidity. The smell of the river was heavy in the air, a mingling of putrid sweetness and lowland fertility.

She told the nun how she had found a husband and child just taken ill, with the mother dying. The nun listened as Constance told the story and watched the manner of her hands as she spoke. "I persuaded her husband to leave the sofa in her room and to go to bed in the next room," Constance said. "While the nurse attended to him, I put the little girl to bed in her crib; she is such a cunning little thing. As I tucked her in, she put one little arm under the pillow and through the bars of the crib and said in the sweetest little voice: 'You can't get that arm under.'" The sister would remember that conversation; it was their last.

A few days later, the ward visitor at St. Mary's pulled Constance aside, his face blanched, to tell her he would not be back the following day. "For I am down," he said. When he reached out his hand, his skin burned her to the touch.

CHAPTER 5

The Destroying Angel

Dr. William Armstrong steered his carriage toward Poplar to St. Mary's Cathedral—he heard the echo of hooves on stamped earth, the rattle of chains and buckles, the horse's bit the accompanying percussion.

Armstrong had been appointed by Dr. Mitchell to oversee the cathedral district near his office on Alabama Street. Though he was used to seeing patients in his office with its cloth-covered rocker and red fainting chair, he had spent very little time there in the past weeks. Healthy physicians were few and far between. Paid ten dollars a day, local doctors and those who came from elsewhere could not make it from one home to the next without being stopped by crowds in the street begging for help. The City Hospital had long since filled its 125 beds, and the doctors now reverted to the days of house calls and saddlebags. Few people would have chosen the hospital over their homes anyway. In the 1870s, hospitals were notorious for spreading disease more often than curing it. People even opted to have surgery performed at home, rather than risk infection in an operating room. Hospitals were essentially for the indigent who could not afford private physicians.

Armstrong continued to live in his own home, but many physicians of the Howard Association stayed at the Peabody Hotel, the only hotel to keep its doors open during the epidemic. After breakfast at the hotel, usually nothing more than bacon, milk and coffee, a doctor was assigned to his particular district, where as many as twenty calls would be waiting as soon as he arrived. The physicians, wearing *Howard Association* armbands, loaded mule carts full of provisions. The doctors carried small leather cases that held knives, scalpels, a spring-loaded bleeding lancet and a pocket watch to take the patient's pulse. With no drugstores open, they carried leaden glass bottles of quinine and arsenic tonic for the fever, as well as vials of ethanol, morphine, caffeine and iodine. They also carried extra handkerchiefs. To identify black vomit, doctors would hold a soiled linen cloth up to the sun and watch the red edge of blood seep from the center like a crimson-colored eclipse.

Physicians reported seeing as many as 100 to 150 patients daily. Their treatments ranged from the practical to the truly bizarre, though all were remarkably similar in their ineffectiveness. Castor oil was given to force the kidneys and intestines to function once again. Sponges soaked in iced whiskey and champagne were used to bring down fevers. Laudanum was prescribed for pain. Citizens also self-medicated—gin sales were higher than ever when a rumor circulated that gin could ward off yellow fever. A doctor was quoted in the *Family Physician* for his treatment of the fever: The patient should sit naked, covered in blankets, on a split-cane, open-bottomed chair above a saucer of burning rum until the vapors caused the patient to faint and fall off the seat.

Dr. Robert Mitchell, however, gave his Howard doctors a specific protocol for treatment. Calomel, an irritant drug, was given to empty the bowels, followed by a mustard footbath and perspiration for twelve to sixteen hours. A sponge bath of whiskey and water followed until the temperature dropped below 102 degrees. Two ten-grain doses of quinine were given, and the patient was to be kept completely quiet—no visitors. Once the fever subsided, a bland diet of milk, limewater or chicken broth followed. No solids for ten days, nor could the patient sit up. Bedpans would be used in the meantime. Hopefully, a healthy family member or nurse could be found to empty them.

As is often the case in *heroic* medicine, the treatment for yellow fever could be as bad as the symptoms themselves. Calomel is mercury based and could cause mercury poisoning if given in the wrong doses or not followed with a saline enema to flush the remaining mercury from the body. And quinine, a derivative of the South American cinchona tree, the *fever tree,* had long been used as a treatment. Unknown to the doctors at that time, quinine is toxic to many bacteria and plasmodium, like in the case of malaria, but has no effect on a virus like yellow fever. Instead, given in high doses, quinine could produce many of the same symptoms as yellow fever: delirium, photosensitivity and nausea.

Though Mitchell desperately needed doctors, he was finally forced to send someone to the train station to turn away volunteers from the North. They did not survive in Memphis but for a few days before becoming patients themselves. The burden was too much.

At night, the physicians gathered to compare notes from the day and perform autopsies in search of clues to the epidemic. The liver, it was recorded, might be the color of boxwood, while the spleen was enlarged and kidneys completely congested. One doctor described the bodies of the freshly dead, which might run temperatures as high as 110 degrees, as having blood that steamed and organs that felt as though dipped in boiling water.

Yellow fever, unlike any other disease, carried a mysterious horror to it. Its attack was acute and quick, its duration painful. In addition to its gruesome symptoms, the fever could cause lacerations and bruises on the skin to openly bleed. Pregnant women spontaneously miscarried. In a testament to the ignorance toward both the fever and women's health, one man wrote that the fever caused women well past their childbearing years to suddenly begin menstruating again. In reality, the hemorrhagic fever led to uterine bleeding just as it did all other types of internal and external hemorrhaging. For the doctors and nurses, the fever's most disturbing symptom must have been the mental decline. In mild cases, it surfaced as irritability and inability to stay still. In severe cases, it bordered on maniacal. Patients ran yellow eyed and delirious into the streets, screamed, thrashed and had to be physically restrained.

William James Armstrong, a thirty-nine-year-old physician, had moved his practice from the country to Memphis in 1873, only a few months before that yellow fever epidemic. New to Memphis, he sent his family away and chose to stay behind in the city during the 1873 epidemic hoping to earn the respect of friends and colleagues. After all, he had a wife and children who depended on his fledgling practice. As a profession in the mid-nineteenth century, medicine was not a lucrative one, nor a highly respected one. In fact, for an educated man with connections, choosing medicine was often seen as throwing away his future. No standard schooling or licensing was required. Most American doctors relied not on science, but on the ability to please patients. In order to build a practice, they established personal, long-standing relationships with families, offering personal advice and treating husbands, wives, children and babies. He might be called in during a complicated childbirth, but even that was handled primarily by midwives and women family members. Physicians in the 1870s had to find a way to remain relevant or necessary to everyday life. When yellow fever struck in 1878, Armstrong again decided to stay in Memphis, sending his wife, Lula, and their eight children to Columbia, Tennessee.

Will Armstrong had a heavy, dark beard and a tender nature. He had married his bride on her sixteenth birthday in the midst of the Civil War, he played the violin and he called his youngest daughter, only a few weeks old, his "dear little pig." As a physician, Armstrong must have seemed gentle, even a little timid.

Though Armstrong had served in earlier epidemics, this one far exceeded the previous ones. He had already lost many friends and acquaintances and feared sending boxes of food, clothes or money from the poisoned city to his family. He wrote to his wife that "the fever is assuming a most fearful form and no signs of abatement. It is not yellow fever such as I treated in 1873. Surely the United States never witnessed such a thing before."

Many doctors like Armstrong had served as physicians or surgeons during the Civil War, but despite the horror of that war, the yellow fever epidemic seemed much worse. The *Boston Medical and Surgical Journal,* in 1878, reported about yellow fever: "It required a much higher order of courage than to risk life on the battlefield, where patriotism, the excitement of conflict and the contagious enthusiasm of masses are a stimulus to noble deeds: these are wanting to the physician who treads wearily along the path marked out by disease and suffering . . . the moanings that ring in his ears are never drowned out by shouts of victory and triumph, and he battles with a foe insidious and unseen till the blow is struck that lays low the victim."

As a Howard doctor, Armstrong spent all of his time on house calls. It was lonely and frightful work, for doctors never knew what they might find when they returned to a house—walls stained with black vomit, delirium, corpses, or worse, patients barely alive, alone and completely lucid. In letters to his wife, Armstrong described the despair settling on him: "I feel sometimes as if my hands were crossed and tied and that I am good for nothing, death coming in upon the sick in spite of all that I can do.

"I never was in all my life," wrote Armstrong, "so full of sympathy and sorrow for suffering humanity . . . God grant that I may be able to administer to the sick throughout."

In September, Armstrong went to visit a friend known as Old Sol (Dr. Soloman P. Green), who lived across the street from St. Mary's. Green had awakened during the night feverish, alone and terrified, and no one heard his cries for help. If taken ill in the night, the doctors knew all too well that no one would find them in their homes. They knew to expect the aches of an approaching fever, the ravaging thirst, the mental decline. And the physicians knew how their bodies, like the dozens they saw each day, would be found as though poached from the inside out. The thought, alone in one's bedroom long after midnight, would certainly terrify the most stoic doctor. As Old Sol told the story to Armstrong the next morning, he wept like a child. "I could do nothing but sympathize," wrote Armstrong.

The sisters at St. Mary's had already promised to find Dr. Armstrong and care for him should he fall feverish alone in the night.

* * *

As days followed nights, there was no measure of time passing, only a blurred sense of sickness and death, of too many cries for help and too few doctors and nurses. Only one change was noticeable among the doctors: the decrease in their numbers.

The cacophony of moans and cries from the ill continued in the halls of St. Mary's. There were not enough sisters to attend, and certainly too few doctors, so only half of the cries went answered. Exhausted, the sisters promised to return to dying patients. More than once, they returned too late or the nuns themselves were found collapsed and feverish in the rooms of patients.

On the last day of August, Will Armstrong was called to St. Mary's on an urgent request. Their dean, George Harris, was down with the fever. He had been without a physician for ten hours, so Constance called for Reverend Charles Parsons to help attend to the dean. When possible, the rules of propriety remained: Male nurses were found for male patients and females for females. Constance told Parsons what to do, how to nurse the feverish patient and together, they waited for Dr. Armstrong. Parsons would also need to take over Harris's duties, for now the nuns would be without their priest.

It was six months since Charles Parsons had stood in full uniform on the eve of Mardi Gras and preached to his Chickasaw Guards; never could he have known he would so soon be in that valley of death he had described, and never could he have foreseen what a hellish place it would be. Parsons had spent every day, sunup to sundown, and well into the night, ministering to the fever victims. Some were parishioners, many were strangers. The service he provided most often was the reading of the last rites, wearing his deep purple stole stitched with a white lily and green leaves, a cross ascending from the center.

Dr. Armstrong fastened his carriage to the post outside St.

Mary's and hurried inside where Constance waited for him. He felt George Harris's feverish skin and studied the languid, depressed countenance of the dean. Armstrong gave the grave news to them that Harris had the fever. He reminded them that it could be a light case, but in a letter he later wrote, "Do not expect to see Dean Harris alive. I worked with him hard last night." Harris would, in fact, recover after a long battle with the fever.

Armstrong quickly packed up his medical case and said goodbye to Constance and Parsons, promising that he would return later. For a moment the three of them stood there together—one who nursed, one who doctored and one who delivered the souls from this purgatorial place.

Charles Parsons found a quiet room and sat down to write a letter to his bishop and friend, Charles Quintard. Sunlight washed the floors of the convent. It was the first day of September, the choking heat showed no signs of relenting, and the death toll rose higher each day. "People constantly send to us, saying, 'Telegraph the situation.' It is impossible. Go and turn the Destroying Angel loose upon a defenseless city; let him smite whom he will, young and old, rich and poor, the feeble and the strong, and as he will, silent, unseen, and unfelt, until his deadly blow is struck; give him for his dreadful harvest all the days and nights from the burning midsummer sun until the last heavy frosts, and then you can form some idea of what Memphis and all this Valley is . . ."

Parsons signed and ended the letter: "I am well, and strong, and hopeful, and I devoutly thank God that I can say that in every letter."

A few years after moving to Memphis, Charles Parsons had remarried; his wife was the niece of Dean George Harris, and during the epidemic she lived with Mrs. Harris on the Annandale Plantation in Madison, Mississippi. In letters to his wife Margaret, Parsons likens the epidemic to the frontline of a battle, in which the firing never ceases. "I never thought I could be happy if you were absent from me but am thankful you are not with me now."

Parsons wrote to Maggie every night of the epidemic, though few letters still exist. One such letter would be found over a century later, part of a package of waterlogged paperwork that survived a fire. It was the last letter he ever wrote: "One of your thoughts, my devoted wife, I know will be that I will have the Fever next . . . I am robust and regular in appetite and sleep, and all that good God Who, in His Infinite Mercy, gave us such a Blessing as you. Kiss my little ones for me. Speak courageous and cheering . . . And God will not forget your labour of love."

The next morning, Charles Parsons awoke feverish. In a warm room, he received a visiting nun from St. Mary's. He was smiling and in good spirits. The nun offered to fan him or hang mosquito netting, anything to make him more comfortable.

"No, no, I beg you will not; indeed, I could not let you so fatigue yourself." The nun looked to the attending nurse who simply shrugged. "Let him have it his way; I never saw anyone so unselfish as he is."

Charles Parsons never descended into the delirium that so often accompanied the disease, and in many cases, was a relief as a patient slipped away unaware of his own suffering or of the family he would leave behind. Parsons continued to talk of his wife Maggie and his "little ones." He remained coherent until the end. In his final hour, he talked of having done his duty, then said he wanted to be taken away from this place. "Where do you wish to go?" he was asked. He signed himself with the cross and mumbled: "We receive this child into the Congregation of Christ's flock and do sign him with the sign of the cross."

On September 6, Reverend Charles Parsons died, and the sisters said, even to the end, he refused to allow any nurses or sisters to waste their time tending to him. He was buried the day following his death at Elmwood Cemetery, and as no clergy were present, Mr. John G. Lonsdale Jr., owner of the private cemetery lot, read the burial service.

The *Appeal*'s editor wrote about his death: "He prepared for it as for battle, and as on a battlefield . . . he fell at his post during duty."

That same day, the *Appeal,* which now only had one editor and one printer left on staff, published another story: "A man on Poplar Street yesterday cowardly deserted his wife and little daughter, both of whom were ill with the fever; if he isn't dead, somebody ought to kill him."

When news of Charles Parsons's death made national news, some thirty priests volunteered to come to Memphis. One was a twenty-six-year-old, idealistic reverend who had just finished services in Hoboken, New Jersey, when he heard of Parsons's illness. A fragile man who had recently recovered from a nervous breakdown, Louis Schuyler was discouraged from going on what would certainly be a death mission, but he was adamant, even stubborn in his intentions: "God calls me. I am safe in His hands—He will do what is best for me."

On the train to Memphis, Schuyler received word that Parsons had died. He arrived in Memphis on Sunday, September 8, one day after Charles Parsons was buried, and went directly to St. Mary's to find Sister Constance and Sister Thecla; the nuns had been without a priest or services for a full week. He was struck with the news that both were down with the fever.

A sister at St. Mary's found Constance resting on the sofa several days before; she was dictating letters and insisted that she was healthy. She had felt the chill come on that morning, but worked for another five hours settling matters, knowing that when she fell many more would follow in her footsteps from neglect and starvation. Constance kept all correspondence, distributed the money, managed what little provisions they had and gave orders to the nuns and nurses.

"It is only a slight headache," Constance persisted when Dr. Armstrong arrived. "I have not the fever, it is only a bad headache; it will go off at sunset." He pulled out his pocket watch to measure her sluggish pulse and stroked his hand against her burning face, then insisted that the nuns give her a cool bath and put her to bed. The sisters made up their finest mattress with fresh linens, but Constance asked for another bed. "It is the only one you have in the house, and if I have the fever, you will have to burn it."

Within the hour, Sister Thecla returned from the deathbed of a patient. Pale and perspiring, she began to shake. "I am sorry, Sister," she said calmly, "but I have the fever. Give me a cup of tea, and then I shall go to bed."

Neither Constance nor Thecla knew of the other's illness, though they lay in rooms next door to one another. Finally, when they kept asking to see the other, the nurses had to tell them the truth: The fever had struck them both and on the same day.

Sister Constance soon slipped into unconsciousness and remained so for most of her illness, waking at one point only to say, "I shall never get up from my bed." By then, 200 new cases of the fever appeared each day in Memphis, and the sister attending Constance wrote, "All the world seemed passing away; the earth sinking from under our feet."

As Dr. Armstrong left St. Mary's late that evening, one of the sisters ran after him and handed him a note. He thanked her and walked out into the night. The carbolic acid dumped into the Gayoso Bayou had killed the fish, and their odor cloaked the neighborhood, burning his eyes. With the sun deep beneath the horizon, the air felt suffocating and the neighborhood deserted. In the distance, two blocks away, the towers and rooflines of the Victorian mansions of Adams Street could be seen like barbed etchings against the indigo sky. When Armstrong returned to his silent house, he lit the lamp and pulled the envelope from his pocket to find a note wrapped around two fifty-dollar bills: "An expression of the affection and gratitude of the sisters." Armstrong sat down at his desk to write his wife, promising that should he survive the epidemic he would repay the sisters. "Sister Constance is dying tonight," he wrote, "and I now think Sister Thecla will get well."

All night the attending sister could hear the moans and delirium from Constance's room. She heard her shout out "Hosanna," and repeat it faintly through the night. At 7:00 the next morning, the toll of the church bell marked the hour. "At that clear sound, which she had always loved, whose call she had never refused to answer," wrote the sister, "the moaning ceased; and at 10 o'clock a.m. her soul entered the Paradise." The chapel was candlelit, the windows streaked with rain. Constance was robed in her habit with roses laid across her breast, a shock of beauty against the gloom. Reverend Louis Schuyler had arrived in Memphis just in time to read the services. Afterward, Constance was taken to Elmwood, where her body had to be held in a borrowed vault, as there were too many dead and not enough gravediggers.

Sister Thecla did recover, becoming a convalescent. Unlike any other disease, yellow fever's hallmark is its cruel tendency to return after a period of brief recovery. When it did, as one doctor warned, it was time to order the coffin. Convalescents were under strict orders to remain in bed and quiet, but nurses and physicians usually hurried back to their duties. The vengeful fever would return with the most severe symptoms. Sister Thecla died one week later, after several days of pain and lucidity. An obituary for the two nuns read, "Of them may it be said that they were lovely in their lives, and in their death they were not divided."

Louis Schuyler returned to St. Mary's after Constance's funeral. He had come to Memphis to fill the vacancy left by dead priests, to offer his services to a congregation of dying nuns and fever patients. Schuyler delivered the news to Dean Harris, who was still recovering, that both Parsons and Constance were gone. "My work here is done," he said, "the whole of Memphis was not worth those two lives." Schuyler left him sobbing.

Schuyler kept no diary or letters, nor would he have had time to write. Or perhaps the terror was too much for the sensitive twenty-six-year-old to record on paper. Even when encouraged to begin slowly, Schuyler had insisted working directly in the neighborhoods hardest hit by the epidemic. He refused a room at the Peabody Hotel for a cot in the parlor of Dean Harris's fever-ridden home. Schuyler was in Memphis only four days before the fever struck him. The beds at St. Mary's were full, and Schuyler was taken to the Court Street Infirmary, which had been recently opened for the feverish nurses and physicians. He was visited by another reverend from St. Mary's, but Schuyler was already wildly delirious. It may have been due to his delirious shouts and screams that Louis Schuyler was moved from his hospital room into the death alley still alive. Piles of corpses and raw pine coffins lay all around him waiting for the wagons, which could take days to arrive. A nurse followed Schuyler's litter into the alley and knelt beside him, promising not to leave his side. They sat beneath the buttressed stone and brick of the alley, cold shadows arching across the skyline creating a mosaic of gray light, sun and blue sky. "Please tell me," asked Schuyler, "whether I am in Memphis or whether I am in my little church in Hoboken?"

On September 11, a cool front brought hope to the city. Rain had fallen the day before, chilling the air and sweeping the bayou clean. Will Armstrong sat at his desk that night at 9:00 writing to his wife, "My heart bounds with joy at the mere hope that this cool night will possibly end our labors . . . No one knows but the weary doctor what a delight that would be. Kiss all the children for me." He ended his letter: "I alone am standing."

A few days later, Lula Armstrong received a telegram from Dr. Mitchell informing her that "Dr. Armstrong is very sick but doing well today. Says you must not come here under any circumstances."

On September 16, Lula received a penny postcard from her husband: "My dear wife: I have passed through the fever stages and have only to get the stomach right. Hope I can do this and see you soon." But by September 20, she was notified by the nuns at St. Mary's that her husband had died of yellow fever. His attending nurse said that even when delirious he tried to rise from his bed to see patients. In Elmwood's leather burial record, the Graveyard Girl recorded his name: *Dr. Wm. J. Armstrong*, wrote ditto marks for yellow fever and the location of his plot in the Fowler Section, Lot #265. His body would be moved years later to another plot where his wife would be buried by his side. Lula Armstrong would also die on September 20 — forty-six years later.

The next day three more sisters at St. Mary's died. The sister who attended Constance at her deathbed soon followed, as did the nurse who attended Charles Parsons. John Lonsdale, who spoke at Parsons's burial, fell feverish and died. John Walsh, the country undertaker, died along with most of his family; at the time of his death, Walsh had buried over 2,000 of the city's yellow fever victims.

Dr. John Erskine, the doctor who opposed quarantine of the city, died on September 17 under the care of his brother, Dr. Alexander Erskine. His death crippled the Memphis Board of Health. It would not begin functioning again until mid-October.

Dr. R. H. Tate, the first black physician to practice in Memphis, was assigned to "Hell's Half Acre" along Lauderdale and Union. He died only three weeks after his arrival.

Three thousand Howard Association nurses, the large majority of them black, served during the epidemic; one-third of those nurses died. Among the 111 Howard doctors, 54 contracted the fever and 33 died.

Charles G. Fisher, head of the Citizen's Relief Committee, died; of the twenty members of his committee, only three were left at the end of the epidemic.

Dr. W. A. White, rector at Calvary Episcopal Church, recovered from the fever just in time to bury his son. A local legend by the name of Annie Cook turned her house of prostitution, the "Mansion House" on Gayoso Street, into a hospital and nursed the sick until she herself perished of the fever. The sheriff died. Even Jefferson Davis Jr., the only son of the Confederate president, was lost to this plague in Memphis. His was the largest funeral seen during the epidemic: Fifteen people attended.

Churches throughout the city sacrificed ministers, priests and nuns. Hundreds more came from cities in the North. Those at St. Mary's have become known as the Martyrs of Memphis.

* * *

At long last, on October 28, a killing frost fell, silvering the tree limbs and blades of grass, cooling the festering quagmire of Happy Hollow. Red leaves littered the ground and gold ones bronzed the treetops. A message was sent to Memphians scattered all over the country to *come home*. That same week, the *Appeal* published a number of advertisements as businesses downtown reopened. Cotton dominated the ads, but a few others touted "New goods at bottom prices," "New mattresses" and "Mourning Goods" like black-trimmed stationery and calling cards, dark cloth and black crepe.

Though yellow fever cases would continue to appear in the pages of Elmwood Cemetery's burial record as late as February 29, the epidemic itself seemed quieted. On November 27, a general citizen's meeting was called at the Greenlaw Opera House. It would be held on Thanksgiving Day, following the holiday church services, to offer the city's thanks to those who had stayed behind to serve and die.

Life was returning to Memphis. Cotton bales began collecting in the streets and along sidewalks. The collective din of steam compressors, train whistles and streetcars could be heard once again. Oyster season had opened, and restaurants and hotels posted signs for "fresh oysters," while Seesel and Son's grocery on the corner of Jefferson and Second received a large shipment of fish. Apples and potatoes filled crates, and mincemeat was prepared. Geese moved south, their wings white with moonlight during the evening hours. Soft rain had fallen early in the week, and men wore their pants tucked up while ladies dragged their hems through the mud downtown. There was even a fresh dusting of snow the day before Thanksgiving, offering a feeling of renewal for some, and for others, just a reminder of the lime that had spent so many weeks on the ground.

Intending to make the Thanksgiving citizen's meeting a taste-

A view of Memphis from the Mississippi River, painted by Henry Lewis in the 1840s.

Memphis/Shelby County History Room, Memphis Library

Cotton piled along the levee in Memphis. In the distance stands the Customs House, designed by famed architect James G. Hill of the United States Treasury. Built in the Italian Villa style of architecture, the Customs House included towers over 100 feet tall, which acted as markers for boat traffic on the Mississippi River. The Customs House was constructed in 1876, but it was not completed for nearly a decade.

Memphis/Shelby County History Room, Memphis Library

Downtown Memphis at the turn of the century.
Memphis/Shelby County History Room, Memphis Library

An advertisement from the Memphis city directory in the 1850s. Bolton, Dickens & Company's main competitor in the slave trade was Nathan Bedford Forrest who, like many other dealers, kept a "mart" on Adams Street.

Memphis/Shelby County History Room, Memphis Library

Invitation to the 1878 Mardi Gras pageants from the secret orders of the Ulks and Memphi.

Pink Palace Museum

Floats from the Ulks parade of fairy tales make their way through downtown Memphis. During the 1870s, the Memphis Mardi Gras parades were the grandest in the country, but after the fever epidemics and the city's financial ruin, Mardi Gras came to a halt in 1882. The modern-day decedent of those parades is the Memphis Cotton Carnival.

Memphis/Shelby County History Room, Memphis Library

The fountain in Court Square was built in 1876 to commemorate America's Centennial. Hebe, the goddess who pours the nectar of the gods, stands atop the fountain, and during the 1878 Mardi Gras, the fountain flowed with champagne.

Memphis/Shelby County History Room, Memphis Library

TENNESSEE — MEMPHIS UNDER QUARANTINE RULE — SISTERS OF CHARITY ADMINISTERING TO SICK AND DYING VICTIMS OF YELLOW FEVER.

Reporters from *Frank Leslie's Newspaper* and *Harper's Weekly* were sent to Memphis during the epidemic to illustrate moving scenes from the yellow fever devastation. Though the illustrations seem cartoonish by today's standards, this was during the time when newspapers offered only text.

Memphis/Shelby County History Room, Memphis Library

Caroline Louise Darling, known as Sister Constance, served as Mother Superior of St. Mary's Episcopal Church. She died September 9, 1878, at the age of thirty-two.

St. Mary's Episcopal Cathedral

ABOVE: The *Emily B. Souder* would have looked much like this ship. Both were screw steamers built in the early 1860s out of oak. Each had three masts, weighed around eight hundred tons and had similar dimensions. The *Souder* was the ship held responsible for bringing yellow fever from Havana to New Orleans, sparking the 1878 epidemic. As the epidemic came to an end in December of that year, the ill-fated *Souder* sank in the Atlantic.

Department of the Navy

LEFT: Charles Carroll Parsons, rector, Grace-St. Lazarus Church, died a martyr on September 6, 1878.

Memphis/Shelby County History Room, Memphis Library

ful event, florists worked for days creating elegant arrangements of azaleas, ferns, begonias, palms and other exotics. The platform of the Greenlaw was grandly outfitted. Colton Greene, the leader of Memphis Carnival who had such hope for the city, was asked to organize the stage decorations. He used that year's Mardi Gras props from the Mystic Memphi.

As the meeting opened at noon, a commemorative statement was made: "To the martyred dead, we feel but cannot express our gratitude; yet, in all days to come shall their memories be kept green, and their names go down in the annals of our city honored, revered and blessed."

Mayor John Flippin, now fully recovered from the fever, had less humble things to say. First he made a statement meant to quiet any gossip and make the record clear for history: At the beginning of the scourge, the press, the city officials and the Board of Health had been true to their promise to proclaim *at once* the appearance of the fever. He followed it with a reprimand for the many who had refused to leave Memphis either from poverty or belief they were immune. "The worthy," he proclaimed, "often perished for the unworthy."

Most important of all, the meeting announced that Memphis, its citizens, representatives in Congress and the Senate would earnestly do all they could to secure passage of a law mandating early quarantine.

In spite of the citizen's meeting and the celebration of Thanksgiving, there remained lasting signs of the plague that November. Schools stayed closed until well into December, and St. Mary's would not open its doors until January. The Greenlaw Opera House, which had once held such promises of sophistication and elegance for Memphis, would be sold as a storehouse by the following spring. Hotels, filled to capacity, promised returning Memphians that their rooms had been thoroughly fumigated and properly ventilated. And Elmwood Cemetery made an announcement that it would allow disinterment and relocation of bodies for the next two months *only*. The Memphis *Avalanche* reported, "Like the Memphis on the Nile, the town was fated to become a ghost city."

December arrived in Memphis, and the worst yellow fever epidemic in United States history ended. That same month, *Emily B. Souder*, the ship held responsible for bringing yellow fever to North America in the spring of 1878, the one that denied fever and instead landed it on the banks of New Orleans, set sail once again for the Caribbean. Somewhere off the coast of New York, on December 10, she sank to the bottom of the Atlantic Ocean, ending her fourteen-year service. The captain and all on board were killed except two castaways.

CHAPTER 6

Greatly Exaggerated

President Rutherford B. Hayes continued to receive word of death and loss at the White House. The 1878 epidemic had stretched from Brazil to Ohio. In the following months, the final death toll in the Mississippi Valley would prove to be 20,000 lives and the financial loss close to $200 million. Two hundred communities in eleven states had been hit by the fever. It was a bitter piece of news. Not only was the 1878 yellow fever epidemic one of the worst disasters to befall the country, but it happened under Hayes's shaky command. "It is impossible to estimate with any approach to accuracy," announced Hayes, "the loss to the country occasioned by this epidemic."

Surely Hayes was also stirred by an unspoken sense of guilt when he thought of his dismissive letter in August; he had called the Memphis pleas for help "greatly exaggerated." The toll on human life in Memphis alone surpassed the Chicago fire, San Francisco earthquake and Johnstown flood combined. It was being called by some the worst urban disaster in American history. Over 5,000 lives were lost in Memphis, nearly a third of the population that remained in August of that year.

President Hayes contemplated his next course of action. Politically speaking, he knew he needed to act authoritatively and quickly. His recent election had been fraught with controversy between the North and South. He had taken office only by assistance from the National Guard and a promise to withdraw federal troops from the South. His liberal views on the rights of blacks further united his enemies in the South. His only allies in that region had been the businessmen and merchants looking to profit from reconstruction, and now, their economic viability would certainly be at stake. On the brink of the 1878 epidemic, the Memphis *Appeal* had published a column entitled, "Poor Hayes: The Republicans hate and mistrust him; and the Democrats, knowing he occupies a position to which another was chosen by the people, have no respect for him." A country not at all secure with his leadership was looking to Hayes for healing.

With Congress out of session in August of 1878, H. Casey Young, the congressman representing Memphis, had appealed to Hayes on behalf of the stricken city not only for relief aid but for a committee of experts to investigate yellow fever once and for all. But the president, one historian wrote, "was not in a position to commit the recessed Forty-fifth Congress to financial sponsorship of an investigation." It would not take long, however, for the epidemic to demand the attention of the U.S. government and launch it into a controversial, public debate over the national health system.

In November, Hayes called his cabinet together to assess the situation; federal response had been slow. In the south, the dead were still rotting unburied in cities and farmlands. Thousands of people had been displaced and collected in camps, waiting for food and supplies. The entire cotton market had been injured. The first action would be to declare a state of emergency in the Mississippi Valley. Charities and state governments could be depended upon to rally support, as well as deliver supplies and wooden coffins. Already, $1 million in aid had gone to Memphis from every state in the union and overseas. Towboats barreled down the Mississippi River carrying over 333,000 pounds of beef, 23,000 pounds of crackers, close to 33,000 pounds of coffee and 200 gallons of whiskey. Train cars full of wooden coffins pulled into the Poplar Street station.

What Hayes needed was a way to comfort the minds of the people, to offer some sense of protection against a tragedy like this in the future. He needed a united force of experts and physicians to not only reassure the people but also actively work to prevent and manage epidemics. The country needed a federal board of health.

In the weeks that followed, a battle would ensue between the American Public Health Association and the Marine Hospital Service, which would later become the Public Health Service. The two agencies, each under the helm of an ambitious, headstrong leader, would fight for dominant control of American health. It also became the familiar battle of federal versus state rights, an echo of pre–Civil War debates. This time, however, southern politicians argued vehemently for federal control of quarantines, while northern owners of those railroads and shipping lines shouted for state control. After all, on a local level, states would be unlikely to risk their own economy with quarantine, much less the financial stability of northern-owned transportation. That very problem had arisen during the 1878 epidemic: New Orleans offi-

cials like Samuel Choppin believed strongly in a quarantine
against infected ships arriving in New Orleans; but, once yellow
fever was present, city officials refused to tell the rest of the country for fear of being quarantined themselves. A reporter for
Harper's wrote in December of that year, "No question in medicine, and scarcely any in theology, has been debated more
learnedly and more ardently—I may say, indeed, more furiously—
nor for a longer time, than this one."

After weeks of investigation, no new medical information was
available, and the country was no closer to solving the yellow fever
mystery. Instead, the entire issue had been overshadowed by par-
tisan politics and backbiting. Only one resolution seemed to
please everyone, and that was the formation of a legislative committee, a joint committee represented by both the Senate and the
House, to investigate the epidemic further. The Board of Experts
would be led by Surgeon General John M. Woodworth, head of
the Marine Hospital Service. Fifteen doctors were appointed to
the board. Several came from New Orleans and other cities in the
South, but also from New York, Philadelphia, Albany and Cincin-
nati. This was a disease, after all, that had at one time or another
affected the whole of the country. Only one doctor had been chosen from Memphis, Tennessee: Dr. Robert Wood Mitchell.

> *Dear Sir:*
>
> *I am authorized by the Committees of the Senate and House of
> Representatives on Epidemic Diseases to advise you that you have been
> elected a member of the Yellow Fever Commission of Experts, with
> compensation at the rate of "ten dollars per day and actual expenses
> while on duty."*

You are respectfully requested to attend a meeting of the Commission to be held in the city of Memphis, Tennessee, on Thursday morning the 26th instant.

Very respectfully,
John M. Woodworth
Washington City, Dec. 19th, 1878

It was December when Mitchell received his telegram, and he was undoubtedly aware of the quarantine debate in Washington, though in Memphis, politics seemed inconsequential to the shell-shocked city. With the dead buried barely a foot beneath the surface, the town still felt like a crypt. Families returned to find homes ransacked, city blocks burned and loved ones buried in the mass graves pockmarking the cemeteries. It seemed the smell of death would never leave Memphis.

Mitchell found little comfort in knowing that he had been right, that an early and efficient quarantine might have prevented over 5,000 deaths. His own Board of Health in Memphis, the one from which he resigned, had been hit severely. Two of its members had been stricken with fever, including the mayor, another board member buried his son, and Dr. John Erskine, Mitchell's nemesis in the struggle, had himself perished of yellow fever. As an army surgeon, Mitchell was precise and objective; he applied his skill to repairing tattered bodies. As the leading obstetrician in Memphis, his work had turned toward delivering new lives. Mitchell's experiences had done nothing to prepare him to watch, unable to help, as so many of the children he delivered, in addition to women and men, friends and colleagues, died.

The first proceedings of the Board of Experts would take place in December 1878, in Memphis, Tennessee. It was a pitifully appropriate setting. The board met on the day after Christmas at the Peabody Hotel at 3:00 p.m. and was there past 7:00 that night. It was resolved that board members would visit the towns most severely hit during the 1878 epidemic, collect blood and tissue samples, investigate weather phenomena and conduct a chemical analysis of the air. Just the idea of testing the air demonstrated the pervasive fear of this disease. In the minds of nineteenth-century scientists, it was as though some unknown, unseen entity traveled through air, climbing, claws extended, into a healthy human to leave a corpse behind.

Mitchell served on the committee to interrogate local doctors and specific cases, as well as quarantine measures. His committee would also keep accurate record of the number of yellow fever cases in 1878, dividing them among whites, blacks and mulattos. The entire Board of Health would meet in Washington, D.C., in January to make presentations.

The findings of these boards have been written into the pages of history in staggering statistics: "Yellow fever should be dealt with as an enemy which imperils life and cripples commerce and industry. To no other great nation of the earth is yellow fever so calamitous as to the United States of America. In a single season more than a hundred thousand of our people were stricken in their homes, and twenty thousand lives sacrificed by this preventable disease."

The board declared that yellow fever made its first appearance in this hemisphere after the discovery of America by Columbus, and it had appeared in a long list of states: Massachusetts, Rhode Island, New Hampshire, Connecticut, New Jersey, Pennsylvania, New York, Delaware, Maryland, Illinois, Missouri, Ohio, Kentucky, Virginia, North Carolina, South Carolina, Georgia, Alabama, Tennessee, Mississippi, Arkansas, Louisiana, Florida and Texas.

Racially, yellow fever decimated members of the white population. While blacks did contract the fever, more so in 1878 than in any other year, the number of deaths was drastically different based on skin color. During the 1878 epidemic in Memphis, the mortality among whites was 70 percent and among blacks 8 percent. In actual numbers, out of 14,000 blacks, 946 died; among 6,000 whites, over 4,000 perished. White Irish immigrants in particular suffered the most. Another unusual aspect of the 1878 epidemic surfaced in New Orleans. In the past, yellow fever had been kinder to children than to adults, often leading to a mild case and a lifelong immunity. That year, New Orleans statistics showed that nearly two-thirds of the deaths were children, the great majority under the age of five.

Most likely, the slave trade had provided a small measure of genetic immunity against the disease for blacks. Living in the South and surviving childhood bouts with the fever also offered immunity. Many of their encounters with yellow fever whether on plantations, in rural areas or in the poor sections of cities most likely went unreported. Whatever the cause of immunity, it had been fuel for racism for decades. White slave owners had argued for keeping slaves as a labor force since they seldom fell to the fevers that so plagued whites in the South.

In the months that followed the Board of Expert's presentation, the argument over public health escalated, and distinct lines were drawn between the two public health giants, the North and the South and theories of how yellow fever spread. Every issue, on all sides, seemed driven by self-interest.

John Shaw Billings, with the U.S. Army Medical Corps, wanted control of the national board, so his American Public Health Association argued that yellow fever was a matter of sanitation, which of course, fell under their jurisdiction. Northern politicians who did not want to see quarantines impede commerce aligned themselves with Billings. If yellow fever was a sanitary matter, why would quarantines be necessary?

On the other side, John Woodworth, surgeon general of the Marine Hospital Service, also vied for control. He wanted strong quarantine powers, which would be controlled by his Marine Hospital Service. It was not the first time, and it would not be the last, that the Army Medical Corps and the Marine Hospital Service found themselves on opposing sides of a yellow fever argument.

The only ones making an honest case in this mire of politics, public insults and self-interest seemed to be the South. They just wanted the yellow fever epidemics to stop, and most southern health officials believed that Woodworth's strong, federal quarantine—whatever the cost to commerce—was the way to do that. Memphis's Casey Young who survived his case of yellow fever that fall argued that he "had fought for four years in trying to make the states greater than the Federal Government, and that effort had cost millions of lives, and this effort made . . . to establish the superiority of the state, if it resulted in defeat of the bill, would result in the loss of many more lives."

Woodworth and his southern supporters lost the debate, and their bill was defeated. John Shaw Billings won the day, and in spite of the ruthlessness of this argument, his place in history remains a great one. Billings was instrumental in opening Johns Hopkins University, and he started the Surgeon General's Library, which would one day become the National Library of Medicine.

Eleven days after the very public debate and humiliating loss, John Woodworth died; his death was rumored to be a suicide. In the wake of this brawl and a divided nation, the National Board of Health was formed.

CHAPTER 7

The Havana Commission

While Memphis struggled to rebuild itself, the nation continued to grapple over the question of what to do about yellow fever. Hayes's Board of Experts had failed to do anything more than provide statistics for past epidemics and the dismal results of the most recent one. What Hayes needed now was a group of experts to go to the source of the problem: Cuba. After all, Cuba had proven to be the hub for all major epidemics of the American plague over the last two centuries. The country could not afford to wait for yellow fever to strike another severe blow. It had to go to the source of the problem and seek out the virus. The National Board of Health organized a group of yellow fever experts to travel to Cuba and study the disease—they were the Havana Yellow Fever Commission.

* * *

Dr. Carlos Finlay had an air of madness about him. He was not mad, quite the opposite; he was brilliant. But he had trouble expressing himself, in part because his mind seemed to work faster than words could accommodate, but mostly because a childhood bout with a nervous system disorder had left him with a distinct stutter.

Juan Carlos Finlay was born in Cuba in 1833. The Finlay family moved to Havana when Carlos, as he would choose to be called, was only one year old. His father was a Scottish physician who was on a British expeditionary force when his ship wrecked near Trinidad, and he met Finlay's mother. In Cuba, Finlay's father practiced medicine and owned a coffee plantation, where Carlos was homeschooled as a child. Finlay's father also loved to travel, and he took Carlos with him on trips throughout the West Indies, South America, and later, Europe.

Finlay's education was multinational as well. He was sent to school in France, but eventually returned after a bout of typhoid. Like his father, Carlos Finlay wanted to practice medicine, but he needed a bachelor of arts degree to do so in Havana. Finlay moved to the United States where medical education was still substandard, and he would not need a degree to enter medical school. He graduated from Jefferson Medical College in Philadelphia.

What Finlay really wanted, however, was to return home to Havana; but, before he could practice medicine there, he would need to pass the oral board examination. Finlay's stutter—paired with Havana's low opinion of American medicine—caused him to fail at his first attempt, but persistence was a hallmark of Finlay's personality. After a year of traveling with his father, Finlay settled back in Havana for good, finally passing his oral boards, and beginning his practice.

Dr. Carlos Finlay was a true intellectual of the Victorian age. He spoke fluent English, French, German and Spanish, and could read Latin; he liked to have breakfast in one language, lunch in a second and dinner in a third. He excelled at chess. Finlay was a member of Havana's Royal Academy of Medical, Physical and Natural Sciences. He was also charitable, often taking on patients who could not afford care. Finlay published articles on subjects varying from cholera to leprosy, gravity to plant diseases, but his most prolific writing involved yellow fever. During his life, he published forty articles on the subject. He was particularly interested in the atmospheric conditions surrounding yellow fever—especially after the 1878 epidemic in the United States. In direct opposition to the prevailing contagionists versus noncontagionists view, Finlay believed that an intermediary host was responsible for the spread of the fever.

In 1879, just after the devastating yellow fever epidemic in the Mississippi Valley and beyond, the group of American yellow fever experts arrived in Havana. The Havana Yellow Fever Commission consisted of several members, including the chairman, Dr. Stanley E. Chaillé of New Orleans, Dr. George M. Sternberg of the U.S. Army Medical Corps and Dr. Juan Guitéras of the Marine Hospital Service. The Spanish government assigned counterparts in Havana to work with the commission, and Dr. Carlos Finlay was a natural choice. Finlay's international background, his congenial nature and his knowledge of tropical diseases made him a perfect fit. One member of the commission would later describe Finlay as "an original, penetrating, tenacious, untiring investigator . . . a mentor worthy of imitation by anyone with a dedicated vocation to science and humanity."

The commission moved into Havana's Hotel San Carlos during their three-month stay. Chaillé was assigned to work on the prevalence of yellow fever in Cuba. Guitéras, a Cuban-born,

American-educated professor of tropical medicine, looked for microorganisms and pathologic changes in the tissue of yellow fever cadavers. And Sternberg searched for a pathogen in the blood samples. Carlos Finlay, Juan Guitéras and George Sternberg would form a lasting friendship during the work—all three would spend the next twenty-five years fighting this disease.

Dr. Sternberg was an expert at photomicroscopy. Using oil immersion objectives and a Tolles amplifier, he produced 105 photographs of blood smears during his months in Cuba. Sternberg, America's "pioneer bacteriologist," also had an impressive résumé. He was captured by Confederates during the Battle of Bull Run, escaped and made his way back to Washington. After the Civil War, he served on the western frontier. During service at Fort Barranacas, Florida, Sternberg contracted yellow fever. He survived the fever, but it launched a twenty-year grudge against the disease he searched tirelessly for beneath the microscope. Like Finlay, Sternberg would publish roughly forty articles on the subject of yellow fever; but Sternberg's expertise was not limited to yellow fever alone. He discovered, the same year as Louis Pasteur, the pathogen responsible for pneumonia, and he was the first in this country to show the malarial parasite and tuberculosis germ. But Sternberg was anxious for his own fame. He was ambitious, and it would take him far. Two decades later, Sternberg and Finlay would again battle yellow fever, one as the most powerful medical mind in America, the other as the most ridiculed scientist in Cuba.

The commission admired and worked well with Finlay, but ultimately were uninterested in his theory about atmospheric conditions, instead focusing on the ever-popular germ theory. They failed to discover any new groundbreaking information on the disease and soon returned to the United States. Finlay's interest in the disease, however, was roused, and he began extensive studies building on the work of the commission. For his part, Finlay was more interested in the hemorrhaging so common to the disease. He believed the "germ" or agent of infection must be spread through the blood. What could pass blood from one person to another? What independent agent could take the blood of one sick person and spread disease to a second one?

There had been some very recent studies on insects as vectors by Patrick Manson, who would later make the connection between mosquitoes and malaria. There was also a French scientist named Louis-Daniel Beauperthuy who had suggested twenty years before that a mosquito—a striped one—had an intrinsic relationship to yellow fever. The fewer mosquitoes, the fewer incidences of fever. Where Beauperthuy missed the mark was in believing that the mosquito just carried filth or decomposing matter, spreading the disease through its bite. Around the same time as Beauperthuy, an American physician, Josiah C. Nott, had also suggested a sort of insect theory, wondering if the yellow fever germ could travel through air much like insects. Beauperthuy saw the mosquito as a vehicle for infected matter; Nott saw infected matter as taking flight like the insect. Both were wrong, but their theories circled the truth nonetheless and broadened thought for future scientists.

To Finlay, the insect theory would also explain why yellow fever epidemics were so sporadic, striking different cities during different years, in spite of quarantines. Finlay was particularly interested in a common striped mosquito, known later as *Aedes aegypti,* which proliferated in areas where yellow fever was present. That particular mosquito had a few peculiar habits that would make it an ideal vector of disease. As soon as it had digested a blood meal, *Aedes aegypti* went in search of another, which would enable it to carry and spread disease easily. The mosquito is also benumbed when the temperature drops below sixty degrees, which correlated with Finlay's atmospheric studies on areas where epidemics arc common and at what times of year they begin and end in places like New Orleans and Memphis. For the first time, it seemed there was a connection between the pest and the pestilence.

It was with a bitter sense of irony that Memphians would one day learn the yellow fever epidemics that nearly destroyed their city, a city named for Memphis, Egypt, would be spread by *Aedes aeygpti:* the Egyptian mosquito.

In 1881, Finlay began studies on *Aedes aegypti* and blood inoculations. His experiments were partially successful, producing a few mild cases. He presented his theory on August 14, 1881, to the Royal Academy under the title *The Mosquito Hypothetically Considered as the Agent of Transmission of Yellow Fever.* To Finlay, the theory made perfect sense, in spite of some inconclusive experiments. But to a medical age wholly dedicated to the germ theory and the idea of contagion, his ideas seemed bizarre. His experiments had also been riddled with problems, leaving more questions than answers. Finlay stood at the lectern and stuttered his way through his presentation, trying to explain his strange theory through fits and starts in his voice. When he finished reading his paper, he looked up and awaited questions from the audience. Instead, he was met with complete silence. The combination of his speech impediment and outlandish theories about mosquitoes left him ridiculed and rejected by the medical community. He was dubbed

"Mosquito Man" by the U.S. press and became known as a "crank" and a "crazy old man" in Havana.

Finlay continued to conduct experiments, including several on Jesuit priests at their monastery outside of Havana. The farm, located high on a plateau in a suburb of Havana, had been leased to the priests for summer residence. In spite of rampant yellow fever epidemics throughout Havana and elsewhere in Cuba, the property, known as Finca San Jose, had never had an epidemic. In coming years, those 150 acres would be a critical setting in the conquest of yellow fever. And for the next two decades, Finlay retreated into his own quiet obsession with mosquitoes.

CHAPTER 8

Reparations

In Memphis, a mild winter approached. Already, people began to speculate that the devastating 1878 epidemic would be followed by another one in 1879. The pattern is consistent with an El Niño cycle, and the loaded mosquitoes from one epidemic would lay virulent eggs that would survive to the next summer, when once again, a yellow fever epidemic would strike.

Another public meeting was called for New Year's Eve, 1878. This time, a handful of prominent Memphians gathered to decide the fate of the dying city. For over a decade Memphis had been bled by corrupt politicians and overwhelming epidemics; it was now $5 million in debt. With a national reputation as one of the most diseased and devastated cities in the country, businesses would surely leave and new ones would not come. Worse, they argued, the flight of so many of Memphis's wealthy whites had left the city with poor blacks and dwindling immigrant populations who could not pay taxes. In the past year the Greenlaw Opera House had been witness to the lavish Mardi Gras party, then a citywide memorial to yellow fever victims, and now, the vote for giving up the city's charter resonated within its walls.

The smell of wet wool and fire smoke laced the dank air of the auditorium. Hoarse voices argued, "Whenever government, from any cause, becomes unable to provide for the peace, safety and general welfare of its inhabitants, it should be abolished and another instituted in its place." Cold hands raised, votes were cast and on the last day of 1878, the Memphis charter was revoked. Memphis became a taxing district of the state; it would remain so until 1893.

The yellow fever epidemic of 1878 altered the fabric of the city forever. By the turn of the century, the original population of Memphis was almost entirely replaced by one much more provincial, Protestant fundamentalism and white supremacy flourished, and cultural diversity all but disappeared. A number of the white Memphians who fled, including the large number of German immigrants and artisans, never returned to Memphis. Immigrant populations ceased to move to Memphis, or to the South in general, in the great numbers they once had. Blacks made up 50 percent of the Memphis population, and the white population consisted mainly of the poor from rural areas of Tennessee and Mississippi. An education census taken in 1918, forty years after the epidemic, showed that less than 2 percent of the white families living in Memphis had been born there. Memphis historian Gerald M. Capers would later write, "It can be suggested with some justification that Atlanta owes its present position as the 'New York of the South' more to the work of the *Aedes aegypti* in Memphis a half century ago than to any other cause."

* * *

In 1879, Hayes's Board of Health hired a New York engineer named George Waring to travel to Memphis and investigate the possibility of a sewer system to clean up the diseased city. Colonel Waring had been to Memphis during the Civil War and was even defeated by Nathan Bedford Forrest in a cavalry campaign near Tupelo, Mississippi. Impressed by a city completely devastated, but still determined, Waring decided he could help: "I had formulated a theoretical system, which had never been put into execution—which probably never would have been put into execution, but for the great needs and the great poverty of Memphis." His idea was relatively simple, involving earthenware pipes sixteen inches beneath the ground, which would carry foul sewage only and exclude all rainwater. The straightforward plan to use separate pipes for sewage and fresh water would become known as the "Waring System." Eighteen miles of sewers, including two main sewer lines along the east and west sides of the Gayoso Bayou, were laid over a period of fifteen weeks. The system proved so successful that cities all over the country soon adopted the design. As Waring himself boasted, "It has attracted attention, and has found some imitators in other countries, and the name of Memphis is known, because of its sewers."

The sewer system was created to clean up the foul, disease-ridden city, but it had another benefit that would not be appreciated for years to come. In eliminating cisterns and providing an efficient means for drainage, Memphis destroyed a large number of breeding places for the striped house mosquito.

Twenty years later, the success of the Memphis sewer system and Waring's work as street commissioner of New York would attract the attention of President William McKinley. George Waring was sent to Havana, Cuba, in 1898, on a sanitary survey for two weeks.

With American troops occupying the island during the Spanish-American War, McKinley wanted the pestilent city clean and disease free. If the Waring system could clean up a city like Memphis, surely it could do the same for Havana. When Waring returned to New York after his survey that October, he felt certain he was up to the task. But within days of returning home, Waring became ill and a physician was summoned.

"I must get up, doctor," Waring complained. "The president is waiting for this report."

"Colonel," the doctor said quietly, "you've got yellow fever." George Waring died twenty-four hours later.

The 1878 epidemic remains a mystery today, as do most other major yellow fever epidemics. What causes the fever to change from an endemic form with a few isolated cases to a full-blown urban epidemic? In 1878, the question might be answered by an extraordinary set of circumstances: The virus may have arrived directly from Africa, which could account for the high number of cases among locals in Brazil and Cuba, as well as blacks in the American South; and an El Niño cycle that year allowed for twice as many mosquitoes in play. In the decade following 1878, the underground slave trade would finally cease in Brazil and Cuba. Sanitation would radically change in cities like New Orleans and Memphis, doing away with the breeding grounds for the mosquito. Fewer immigrants would venture south. Though New Orleans would suffer another epidemic in 1905, it would never travel farther north. And so, 1878 remains the last great epidemic of the American plague on North American soil.

Amid the devastation in the South and the long road toward rebuilding it, Americans barely noticed another significant event of 1878: Cuba finally lost her fight for independence from Spain at the close of the Ten Years' War. American politicians turned their attention to the tropical island, a major shipping port and supplier of sugar.

In 1898, just twenty years after the crushing yellow fever epidemic, America would find herself in Cuba to fight an old enemy, but it wasn't Spain; it was yellow fever.

PART THREE

Cuba, 1900

The prayer that has been mine for twenty or more years,
that I might be permitted in some way or sometime to do
something to alleviate human suffering, has been answered!

—WALTER REED, December 31, 1900

Cuba, 1990

CHAPTER 9

A Splendid Little War

On a still and starless night, Captain Charles Sigsbee felt the ship beneath him shudder.

He had just taken a seat at his wooden desk, straightened a piece of paper and begun to write a letter to his wife. It was a warm night, and the cabin mess attendant had delivered Sigsbee's civilian's thin coat to wear in the heat. As the captain reached into his pocket, he pulled out an unopened letter, dated ten months ago, and addressed to his wife by a friend. Sigsbee sat down to write an apology for forgetting to deliver the note; he wrote the date at the top of the letter, "February 15, 1898."

The sorrowful notes of taps penetrated the metallic walls of the ship to the quarters below. "I laid down my pen," he later wrote, "to listen to the notes of the bugle, which were singularly beautiful in the oppressive stillness of the night."

Most of his enlisted men had already fallen asleep, rocking in their hammocks, suspended like cocoons between the heavy beams of the ship. To those on the shore, the ship appeared as a great garrison of steel, measuring 319 feet, 6,683 tons and housing 328 souls. Two massive smokestacks towered over the decks, and in the scattered moonlight, water sparked as it lapped against the hull.

As Sigsbee folded his letter and slid it into the envelope, he felt the ship rise as though an enormous ocean swell had passed beneath her; then he heard the inhuman wail of twisting metal followed by the sound of screaming men. The lights went out on the USS *Maine*.

On the seafront of Havana, in homes and cafés, the explosion from the harbor shook furniture, shattered windows and un-hinged doors. Every light in the city went dark, and people ran into the streets, drawn to the show of rockets and fireworks. De-bris flew 150 feet in the air, raining paper and fragments over the ship. Gray smoke billowed in the coal-black sky, while orange flames licked below. They watched as one of the twin smokestacks of the *Maine* heaved over, and the bow disappeared into the black-ness as fire consumed the ship. The reflection of the inferno on the water turned the harbor a glaring red.

Sigsbee groped his way through the darkened, smoke-filled quarters, feeling the *Maine* rolling seaward. Squinting and trying to adjust his eyes to the dark, he took in all the damage around him. The explosion had taken place near the front ammunition magazine in the forward part of the ship—right below the berth where the enlisted men slept in their hammocks. "On the white paint of the ceiling was the impression of two human bodies— mere dust," he would later report.

As the captain made his way onto the deck, a dreadful calm and discipline prevailed in spite of all the violence. The captain was informed that the explosion had taken place at 9:40 p.m. The ship had sustained much damage, and one of the smokestacks was lying starboard. The compartments below were filling with water, and the *Maine* would soon go under.

Cries from men in the water echoed: "Help! Lord God, help us! Help! Help!" Sigsbee ordered his officers, most of whom had been spared in their quarters far from the explosion, to lower all undamaged lifeboats and set out to rescue the sailors of the *Maine*. Boats from the *City of Washington*, as well as Spanish seamen, paddled toward the wreckage to rescue the wounded. The *Maine* continued to sputter and rupture as flames ignited live rounds aboard the ship until nearly 2:00 a.m.

Newspapers would write that Captain Charles Sigsbee had heroically stayed on board the *Maine* until he was the last man to leave his ship. Sigsbee saw it differently: "It is a fact that I was the last to leave, which was only proper; that is to say, it would have been improper otherwise; but virtually all left last."

The wounded were taken aboard nearby ships or carted to the hospitals on the shore. Sigsbee worried that his sailors would be taken to hospitals where yellow fever existed, but there were no hospitals in Havana that didn't house the dreaded fever. Doctors, nurses and civilians tried to mend the crushed bones, deep cuts and hideous burns of the sailors. Many survived, but only partially, losing eyes or limbs or faces in the process. When the final death toll came in, including a number of wounded who later died, 268 had perished in the explosion.

Aboard the *City of Washington*, Captain Charles Sigsbee dispatched his telegram to the secretary of the navy in Washington, D.C. He ended the message with a warning: "Public opinion should be suspended until further report . . . Many Spanish officers, including representatives of General Blanco, now with us to express sympathy."

Regardless, two days after the tragedy of the *Maine*, William

Randolph Hearst sent forth his morning edition with a definitive yet unsubstantiated headline: "Destruction of the Warship *Maine* Was the Work of an Enemy." The enemy, the paper clearly illustrated, was the Spanish.

A Spanish officer in Havana held the newspaper up to Charles Sigsbee. In it, he could see an artistic rendering of his ship, as she once was, anchored above a Spanish mine. In another illustration, wires connected the *Maine* to the Havana shore.

"What do you think of that?" the Spanish officer asked.

Still irritated that the Havana newspapers had been unfair to him in the past, Sigsbee remarked, "If the American newspapers gave more than the news, the Spanish newspapers gave less than the news: It was a question of choice."

Tensions in Havana flared over the next few days as separate American and Spanish inquiries studied the remains of the *Maine*. Sigsbee watched helplessly as divers picked through his capsized ship and bodies were recovered. More was at stake than just the destruction of a U.S. Navy vessel and loss of seamen.

The apex of all shipping that came into the Caribbean and an island rich in sugar plantations and workers, Cuba had long been considered a piece of prime real estate for expanding America. At least four U.S. presidents had attempted to buy it. John Quincy Adams had called it a natural appendage to North America, and Thomas Jefferson believed it to be "the most interesting addition which could ever be made to our system of States." Before the Civil War, the South hoped to annex Cuba as a slave state; after the Civil War, the North looked to it as a source of raw materials. The latest proponents including Theodore Roosevelt, Henry Cabot Lodge and Henry Adams, a group John Hay dubbed the "Pleasant Gang," considered Cuba part of our great manifest destiny. They quoted the likes of Charles Darwin, Rudyard Kipling, Frederick Jackson Turner, and amid the cigar smoke in Washington salons, it seemed clear that America had not only a right but a duty, bolstered by a growing navy, to enlighten others and protect our interests in the western hemisphere. America had sprawled westward, settling the Atlantic coast to the Pacific, and it had grown restless. Cuba was the new frontier. Not all among their ranks subscribed to the idea of expansionism, but if they disagreed with it, most did so out of another sort of elitism: They feared muddying American waters with other races, uneducated cultures or sickly immigrants.

In fact, the issue of disease had become yet another facet of imperialism. Bacteriology had been dominated by Germany and France, but tropical medicine became medicine's new frontier for England and the United States. Colonizing Africa, India and the Americas would be impossible without controlling the fevers so deadly to white settlers. What's more, medicine itself began to take on an imperialist slant—if disease spawned and spread from tropical countries like Cuba, it was America's responsibility to step in and clean them up for the sake of American health. As Robert Desowitz, a professor of tropical medicine, wrote: "An undercurrent of opinion had long held that the United States should take over Cuba for medical reasons, a yellow fever cleansing. The sinking of the *Maine* was just the ticket to do so."

In this spirit, Americans vehemently debated the question of Cuba over dinner tables, in men's clubs, even from the pulpit. News of Spanish concentration camps and starving Cuban prisoners softened American sentiment toward intervention, the prospect of sugar softened economic reasoning, and the thought of toppling a smug European presence just seemed appealing in general.

As the *Maine*—and national pride—erupted in flames and

sank into a watery grave in the Havana harbor, America had its battle cry. The cause of the explosion would be publicly debated in the weeks following the tragedy, and well into the next century. A 1976 examination of the USS *Maine*'s records finally resolved that the explosion, an internal one, most likely resulted from spontaneous combustion: The coal had been located dangerously close to the ammunitions magazine. In the three years before the *Maine* steamed into the Cuban port, a number of other vessels reported fires in the coalbunker. In 1898, however, the United States Court of Inquiry found a different cause: An enemy mine situated under the bottom of the ship.

If the reason for war with Cuba was not immediately evident, it soon found its clarity in the publications of William Randolph Hearst and Joseph Pulitzer. In the end, the Spanish-American War would be the most popular war this country ever fought. While only 1 in 6 soldiers would make it into combat, 200,000 volunteers boarded trains and waited in American military camps hoping to go. The war would be won in just 113 days, liberating Cuba, adjoining Puerto Rico, Guam and the Philippines to the United States.

In spite of this, the war would cost more lives than ever expected. Over 2,500 American soldiers would be lost—not to the Spanish, but to disease. Only 385 would actually die in action. With far more volunteers than there were accommodations to hold them, soldiers crowded into American camps hoping to make it to Cuba before the fighting ended or before typhoid or dysentery picked them off one by one.

Conditions in Cuba were no better, where not only typhoid, dysentery and malaria ran rampant but also yellow fever. As fever season encroached, one soldier wrote that taps was played continuously in the camp: "The volleys became more frequent and one bugle followed another throughout the day; they followed each other almost as if they were but echoes among the hills about us." Eventually, the camps had to stop playing taps for dead soldiers because the unceasing bugle notes brought down general morale.

CHAPTER 10

Siboney

The rains started in Cuba, falling lightly at first, then in heavy, gray panels, drumming against the blades of palmettos and muddying the camp. In the afternoon, the temperature might reach 120 degrees, but during the night, it dropped into the 60s, and rain fell steadily, soaking the soldiers in their dog tents. Camped outside the village of Siboney on a narrow mesa between steep hillside and jagged, coral-rock shoreline, the Thirty-third Michigan and Seventh Massachusetts awaited the Battle of Santiago.

Dr. Victor Clarence Vaughan, dean of the medical school in Michigan, chose to sleep in a hammock. It was a good way to prevent the crabs that scratched their way across the sparse mesa from crawling over him in his sleep. With his poncho and blanket hanging in the boughs of the tree to protect against the rain, he could reach the blanket when the temperature fell and wrap up inside the wet wool while waiting for morning and the swarms of mosquitoes that came with it. When the mosquitoes were too much, Vaughan would strip down and head toward the water pipe that pumped fresh water from the mountains outside of Siboney into Santiago. The troops would drive holes into the pipeline to drink water, or even better, get a morning shower, before plugging them closed again. After his pipeline shower, Vaughan would make his way to the mess hall for coffee—freshly made by the cook, who pounded the grains in a tin pan with a bayonet.

The army needed a place to unload supplies, but the treacherous road from the American-held Guantanamo Bay to Santiago was riddled with tangled jungle and tropical fever, so transports— including a large number of medical supplies—were sent to the coastal hamlet of Siboney under the command of General W. Rufus Shafter. Vaughan and the other Michigan volunteers had arrived there on board the *Yale* on June 27, 1898.

Vaughan had already seen things in Cuba he never expected to see—some were innocent observations like the fact that the locals feared the soldiers' toothbrushes. The soldiers kept them in the bands of their caps, and the locals, having never seen a toothbrush before, thought it might be some sort of weapon. Other observations were more disturbing. The Cuban insurgents, so romanticized in the U.S. as revolutionaries, were nothing more than a collection of withering men, women and children. "It was the first time I had seen a starving people, and I could hardly believe what I saw," wrote Vaughan. "My mental picture of starving people had consisted of individuals uniformly emaciated from head to foot, but the first impression made upon me by these people was that they had eaten too heavily. The limbs and the chest were greatly emaciated, while the abdomen was markedly protruding."

* * *

Vaughan and the troops were at their new camp less than a week before the Battle of Santiago. A few miles away, Teddy Roosevelt and his Rough Riders rode toward Kettle Hill in the San Juan Heights. In the next six days, 12 surgeons—Vaughan among them—would treat 1,600 wounded men from the battle. They operated throughout the day and by lanterns swinging from posts in the night, listening to the sound of Spanish bugles in the distance and smelling smoke rising from the sea. Since most of the bullets and shrapnel passed directly through the body, the surgery was fairly simple. Vaughan chose a barber as his surgical assistant to help him clean wounds with Lysol soap and apply iodoform gauze at the flesh points of entry and exit.

When all of the men had been treated and the wounds dressed, Vaughan went back to his hammock exhausted, convinced he could sleep anywhere under any conditions. Exotic birds called from the treetops, and the smell of rotting mango and approaching rain was in the air. Cloudbursts would fall by the afternoon. As he reached his hammock, another stretcher arrived. The patient was very sick, but sleeping soundly at the moment, so Vaughan asked the men to place the stretcher underneath his hammock where the patient would be shaded while he slept, and then Dr. Vaughan drifted off to sleep.

He awoke to the sound of retching.

Leaning over the side of his hammock, Vaughan saw ink stains of black vomit on the man's face, clothing and blankets. He sprang out of the hammock and ran into the camp.

"Come, doctor, we have a case of yellow fever," Vaughan shouted at Dr. Guitéras, the well-known yellow fever expert and professor who had served on the 1879 Yellow Fever Commission.

They did what they could for the fever patient, then quickly set out to find a site for a yellow fever hospital, far from the other sick wards. Their haste was critical in the case of a yellow fever outbreak.

Guitéras and Vaughan rode a handcar up the mountain railroad east of Siboney. The place they chose was more than a hundred feet above sea level along the northern slope of the mountain range overlooking a huge valley and the Sierras beyond. For reasons unknown to them, high ground seemed the best place to treat and prevent the spread of yellow fever. Tents and supplies were set up immediately, and before nightfall, there were already three patients in the new hospital.

General Shafter wrote to Washington, "There are now three cases of yellow fever at Siboney, in Michigan regiment, and if it gets started no one knows where it will stop." That number grew to thirty the next day, then hundreds. By the end of the epidemic, close to fourteen hundred cases would be seen in the makeshift hospital overlooking a valley of royal palm trees and mountain peaks; the village of Siboney would be evacuated and burned to the ground.

The epidemic presented another problem: American soldiers could not return home until the fever could be contained. Instead they waited, like men anticipating execution, as yellow jack moved through the troops. One all-black regiment was ordered to Siboney to care for the fever patients; by the time they left, they had lost one-third of their men. Yellow fever quarantine camps had been set up in New England to accommodate returning soldiers, but a handful of senators from that area called the war department with objections. It was decided to keep the soldiers in

Cuba until they could be certain yellow fever could be controlled. Even letters out of Santiago had to be thoroughly fumigated in the U.S. before delivery.

In what would be known as the "Round Robin" letter, General W. R. Shafter and Colonel Theodore Roosevelt devised a plan to write a letter signed by the officers under Shafter's command that would be leaked to the press to force public opinion. Public pressure would surely drive McKinley to bring his troops home rather than leave them to the ravages of yellow fever. Although typhoid and dysentery had been far more damaging to the troops, the officers feared the onset of the fever season would be disastrous for the already sickly troops. "All of us are certain," wrote Roosevelt, "that as soon as the authorities at Washington fully appreciate the condition of the army, we shall be sent home. If we are kept here it will in all human possibility mean an appalling disaster, for the surgeons here estimate that over half the army, if kept here during the sickly season, will die."

Roosevelt's letter as well as the Round Robin letter signed by Roosevelt and Shafter's other officers were given to a reporter from the Associated Press. The plan worked beautifully, and the next day, the *New York Times* headlines blared: "War Department Spurred to Activity by News from Santiago: Col. Roosevelt Declares 90 Per Cent of the Army Is Incapacitated and That Men Will Die Like Sheep If Left in Cuba."

The situation had been gravely misrepresented from the beginning. Shafter had neglected to give accurate reports to Surgeon General George Sternberg and Secretary of War Russell Alger. Most likely, Shafter was fearful of news of sick soldiers spreading to the Spanish; but he also worried that camp surgeons were confusing cases of typhoid and malaria for yellow fever. The Round Robin letter published by the press came only three days after

Sternberg and Alger first knew the full extent of the epidemic spreading among the Fifth Army Corps.

The administration was furious. McKinley was in the midst of negotiations for peace with Spain, and the Round Robin letter, he argued, damaged his position. Why would Spain negotiate if they knew biding their time during yellow fever season would wipe out half of the U.S. Army in Cuba? But the Spanish undoubtedly had their own concerns to contend with; in the previous four years, 16,000 of their troops had been stricken by yellow jack. In the end, the Round Robin controversy was a moot point. Orders had already been placed, even before the publication of the letter, to return the troops to the U.S.

Troops were sent to Montauk Point, a cool, high ground, to wait out the yellow fever quarantine. In all, 20,000 soldiers would land at Montauk Point in the space of three weeks. Conditions at that camp and elsewhere were notoriously bad. Surgeon General Sternberg had warned the administration that disease would be devastating to the army; he had been right.

Bald, with a white walrus moustache, Sternberg was a lifelong army man: straight backed, square shouldered, stout, a look of purpose in his face. He was a product of Lutheran ideology and German descendents. He was described by some colleagues as stiff and even egotistical, but he had real ambition, and it had taken him far. After his service on the Havana Yellow Fever Commission in 1879, Sternberg had continued his work on yellow fever, searching for the pathogen in the blood that caused the fever. He had even put forth one theory that bacteria he discovered, called Bacillus X, were the culprits. By 1893, Sternberg had been named surgeon general, and he opened the Army Medical School in Washington, D.C.

But that ambition had also led to some disappointments. As
"America's Bacteriologist," Sternberg was a forward-thinking man
who had made important discoveries, but not before the great suc-
cesses of France's Louis Pasteur and Germany's Robert Koch. His
place in medical history was in no way secure, and what he needed
now was his own definitive discovery.

George Sternberg was in a tough position as the Spanish-
American War escalated. The military was wary of medical officers, seeing them most often as obstacles, or even annoyances,
during war. As the surgeon general, the medical officers beneath
him often saw Sternberg as a part of beaurocratic machinery. An
editorial in the *New York Times* accused Sternberg of lacking the
nerve and force of character necessary for a wartime authority figure. Another *New York Times* writer remarked, "Surgeon General
Sternberg is unfit for the position he holds and that to his ineffi-
ciency is chiefly due the complete breakdown of the medical department in this war."

Nonetheless, Sternberg had advised against sending troops to
Cuba during the wet season when yellow fever was predominant,
and he went so far as to refuse sending nonimmune army medical
officers into fever-ridden Cuba. Sternberg also published concerns for the unsanitary conditions of camps and rapid enlistment
of nonimmune men. There were simply too many men crammed
into too many camps to keep up any sort of acceptable hygiene.

One week after the first case at Siboney appeared, Dr. Vaughan
was walking the grounds of the camp when he felt a severe pain in
the small of his back—he could barely stand or walk, and he knew
immediately what was happening. He limped back to his tent to
write a letter to his wife. His guilt mounted as he put pen to paper
and wrote that he had been ordered into the interior of Cuba for

the next two weeks, and she would not hear from him. Surely the lie would be more comforting than no word at all, or worse, the truth. After that, Vaughan went to see the chief correspondent of the Associated Press, asking him not to mention his name or his condition in dispatches home. Then, Vaughan entered the white hospital tent that had housed surgical patients: Long rows of empty cots, over fifty of them, spread out before him in two straight columns. He asked an orderly to change the sheets on one and fell into it.

That night, Dr. Guitéras examined Vaughan. "Only a little malaria. You will be all right in a few days. Tomorrow I shall give you quinine." Guitéras told the nurses to keep Vaughan comfortable and see that he has everything he needs; then he left.

Throughout the night, Dr. Guitéras returned to Vaughan's bedside—half a dozen times. Vaughan kept his eyes closed and pretended to sleep each time Guitéras entered, but he could always feel the doctor's touch on his wrist and hand on his brow.

As Guitéras came into the hospital tent the next morning, he was swinging his arms and whistling a tune; he gave the impression of a man who had a long night's sleep and nothing in the world to worry about.

"Only a little malaria," Guitéras repeated. "Don't you think that the air up at the yellow fever hospital on the mountain side is much better than it is down here on this low wet ground?" Defeated, Vaughan said that he was ready to go to the yellow fever hospital and tried to rise from his cot, but Guitéras gently pressed him back down into the bed.

"Your temperature is above one hundred five degrees; your pulse is below forty; a change in position, even the sudden lifting of an arm, might stop your heart. You will not move on any account. Men will come, lift your cot, place it on a flat car, and you will be carried to the yellow fever hospital. I shall go with you."

Vaughan spent the next week in the yellow fever ward, his stomach painfully contracting and shrinking, only to suddenly burst again with black vomit. He was treated with calomel and a local lemonade made from Epsom salt, lime juice and warm water. One afternoon, another yellow fever patient offered Vaughan a ginger ale. The thought of the carbonated drink was overwhelmingly appealing, but Vaughan cautioned the man to wait another few days. The patient insisted, trying to rise from the bed to find a corkscrew for the bottle.

"Do not move," Vaughan whispered. "If you must drink the ginger ale, call the orderly and have him pull the stopper. Your heart is crippled and a change in position may kill you."

The man laughed, rising from the bed, and then fell dead across Vaughan, breaking the cot beneath them. Vaughan called for the orderly, and the body was removed.

Another day, a yellow fever patient bribed an orderly to bring him solid food when he was under orders not to eat. He died within a few hours of the meal. Though his hunger grew intense, Vaughan adhered to his liquid diet of warm lemonade. He also suffered from delusions, believing that he was at once the patient and at other times just observing the patient—a "double consciousness," he called it.

His temperature climbed toward 106, and he watched the clouds gather over the Sierras. He began to see gods and demons standing on the mountaintop. The clouds grew wilder and thicker, eclipsing the sun and swallowing the world. Each day, those lightning storms and clouds gathered in the mind of Victor Vaughan, until finally, they parted and disappeared suddenly and completely. Vaughan's fever had broken, though he was now sixty pounds lighter and considerably weaker.

His superior officer prepared orders to send him home by transport, but Vaughan refused, arguing that his new immune status would enable him to do even greater work in the hospitals in Cuba. Rather than simply order Vaughan to return, the officer instead told Vaughan to stay as long as he liked and set up a tent for him—just a few feet away from the mess hall. Vaughan could smell the food from the mess, and several times he attempted to get out of his cot and walk there. He was too weak to manage even the short distance and had to crawl on his hands and knees. Vaughan's superior officer visited him every hour.

"Tomorrow morning a transport leaves for the United States with convalescent soldiers, and I haven't a doctor to send with them," complained the officer. "I do not know what to do." It was a thoughtful and effective tactic.

"Major, do you want me to go on that transport?" Vaughan asked.

Without answering, the superior officer called to a captain, "Bring a stretcher with bearers and put Vaughan on the transport." On the long journey home, Vaughan recovered fully from his bout with yellow fever.

Victor Vaughan arrived back in New York in August of 1898, where orders from Surgeon General Sternberg awaited him. He would be part of the Typhoid Commission to investigate disease rampant among the American camps, and the head of the board would be Major Walter Reed.

CHAPTER 11

An Unlikely Hero

Walter Reed wore immortality modestly. He had a moustache, long and ribbonlike, on an otherwise boyish face. He referred to his wife and daughter with gushing pet names and had a habit of rubbing his palms together when pleased about something. His favorite drink was mint julep, though he was a minister's son who could recite Scripture flawlessly. His lanky build belied a posture spent crooked over a microscope. His narrow, gray eyes were earnest, his brow creased with age. He looked more like a physician than a soldier, which is probably how he liked it since he preferred the men under his command to call him doctor, rather than major. Still, Reed was the type of army man and physician always in uniform and always within code; throughout his life he would write the word *duty* with a capital D. His favorite poet was Sir Walter Scott, the poet of chivalry and honor. If any pride or egotism existed within him, it did so far beneath an exterior of humility—his brother Christopher once remarked about him that every modest man is not great, but it is equally true that every great man is modest.

That Walter Reed's name would survive among the greatest names in medical history would certainly have come as a shock to him had he lived long enough to learn of it.

Reed was born September 13, 1851, in a small, milk-white cabin in Gloucester County, Virginia, to a Methodist minister named Lemuel Sutton Reed and a mother named Pharaba White Reed. The cabin was on loan to the family after the parsonage burned shortly before their arrival, so Pharaba gave birth to her youngest child in a two-room cabin cradled by an elm tree.

Both Lemuel and Pharaba came from colonial families out of North Carolina, their heritage rich in self-reliance, inventiveness and fairness, traits that were fast becoming the trademarks of American ingenuity and success. In this new world, there was no sense of entitlement or aristocracy; determination and hard work were the blueprints for achievement. Pharaba's finest attributes surfaced in Reed, even at a young age. Aside from her fair hair and blue eyes, Reed inherited her intelligence, vivacity, sharp wit and a love of gardening. He would follow his mother around the garden, imitating her, as she pruned the cornflowers, roses and larkspur. It was a passionate hobby he carried with him during his nomadic years in the army.

As a "circuit rider," Reed's father moved every two years throughout Virginia, wherever Methodist churches were in need of a new minister. Ministers of small towns maintained a certain celebrity and power over parishioners, and with it came social responsibility. Towns would wait to see if the new minister would tolerate dancing, drinking. Would he be a strong speaker? Would his family set the right example? Receiving very little pay, ministers and their families relied heavily upon their philanthropic neighbors. Reed may not have taken so well to the evangelical lifestyle though; he would not officially join the Methodist Church until he was a teenager. He joined the day his mother died. And, years later, when his son, Lawrence, was in school, Reed was incensed to learn of a Methodist revival there: "I don't approve of such things for Children," he wrote. "My boy hasn't done anything that he should be told that he is lost since and in danger of hell-fire."

During the 1850s, Virginia breathed promise in her verdant landscape, still largely made up of tightly knit farm communities. For a young boy, there could be little else to worry about but hiding among the languid leaves of tobacco plants, playing with siblings, and studies in the one-room schoolhouse. Then war came.

As "Lee's Miserables," as the troops came to be known, waged a losing war, Union soldiers swarmed into Virginia, toward Richmond. The sound of cannons trembled in the night air, smoke garlanding above the farmland. For Virginians, it meant food shortages, rationing, ruined farmland, failed crops. War pierced the general calm of the Reed household as well when Reed's two oldest brothers enlisted. At Antietam, a cannonball struck his brother James. In a crude hospital tent, the surgeon amputated James's hand. He returned home like so many of his peers, physically clipped and psychologically broken.

Though the battlefield proved treacherous during the Civil War, disease was the greatest enemy. An estimated two-thirds of Union and Confederate soldiers died of disease, not wounds. Yellow fever was one of the most deadly. With this in mind, one of history's more famous examples of germ warfare surfaced. Dr. Luke Pryor Blackburn of Kentucky, known as "Dr. Black Vomit," attempted to ship infected clothing from yellow fever victims to major northern cities. He also had plans to poison New York's water supply, as well as burn the city. He then turned his attention to assassination, sending trunks of clothes from the yellow fever wards to President Lincoln in hopes of giving him a lethal dose of the fever. Again, his attempts were unsuccessful thanks to ignorance about how the disease was spread. His criminal efforts did little to harm his career, however. After the war, Dr. Blackburn continued to practice medicine and aided victims during the 1878 epidemic in Memphis. In the following years, he would become the governor of Kentucky.

A good student, Reed attended school every year except one during the height of the Civil War. He drank knowledge. Not born to privilege, he was equipped instead with other tools for success—lofty principles and swelling determination. As his older brothers finished their schooling, Reed's father asked the Methodist Church if he might move to Charlottesville to minister and allow his boys to attend the university.

Walter Reed was fourteen when the family moved to Charlottesville, on the cusp of the Shenandoah Mountains. Charlottesville rattled with traffic along the cobblestone streets. Churches and storefronts lined the walks. The town even had its own daily paper. It was here that the Reed brothers entered Mr. Jefferson's university; the school made an exception in allowing Walter Reed to begin at age fifteen because his elder brothers were also in attendance.

As a boy, Reed was thin boned, determination and conviction visible on his face. His cheekbones were high, his eyes set close to one another and his brow narrow, so that he had a look of steadfast concern that would stay with him well into adulthood. But it was not his physical appearance that people noticed. People took to his gracious and well-mannered nature. He was honest and likable, and that left a marked impression.

Reed appeared before the board at the University of Virginia with an erect posture, chin set forward, trying to look older than his age. Reed had two older brothers at the university, and his father would not be able to afford the remainder of Walter Reed's education. To earn a master of arts degree, a student completed studies in eight different departments ranging from Latin and Greek to mathematics. Reed had only finished courses in Latin, Greek and history and literature—he was far short of the requirements for an M.A. degree. Reed asked permission to attain a degree in medicine instead. True to the time period, an M.D. could be earned more easily and in less time than an M.A. The board considered his request impossible, so they gave their consent. They made a gentleman's agreement on the deal. Should this boy with only one year of university experience pass the medical examination, the school would allow him to graduate with a degree in medicine.

Reed studied for nine months, taking courses in anatomy, chemistry, pharmacy, surgery, among others, and slept only three or four hours a night. At the end of the term, he passed the examination, and the board kept their promise. Walter Reed remains the youngest graduate of the University of Virginia's medical school—he was seventeen years old.

Following his graduation, Walter Reed moved to Brooklyn with his brother, Christopher, an attorney who wanted to practice law in New York where he would never run out of clients. For Reed, New York represented the best in medicine—its hospitals highly regarded and its many citizens in need of help. It was not, however, an easy choice for either brother to make as southerners. A mere four years since a war still bitterly fresh in the minds of both sides, the North felt like enemy territory.

Reed earned his second medical degree from Bellevue Hospital Medical College, the premier teaching hospital at the time. Because he was only eighteen years old, Bellevue withheld his degree until he was twenty-one. At Bellevue, Reed had his first opportunity to use a microscope, as well as a thermometer—measuring ten inches long and taking five minutes for a reading. Medical studies at that time were still very crude. Cadavers were not preserved well, so doctors often kept handkerchiefs over their mouths while working. The stethoscope had been around for only fifty years. Disease prevention was unheard of. Links between sanitation and illness had not yet been realized.

Reed continued to practice medicine in Brooklyn and New York, as well as serving on the Brooklyn Board of Health as a sanitary officer. New York City teemed with immigrants after the Civil War. For the first time, Reed met Italians, Irish, Asians, Germans. With the mass immigration came new epidemics, which spread rapidly among the conditions of filth and poor nutrition in the city. Reed was assigned as district physician to the poorest district in New York. Many of his patients included the destitute "squatters" who lived in colonies in Central Park.

Growing up in Virginia, Reed lived mostly among farms, his family dependent upon the goodwill of parishioners for extra provisions. Neighbors helped one another. Never had he seen real urban poverty and the self-serving individualism that accompanied it. He was appalled by the unsanitary living conditions and poor medical attention given to lower classes in the cities. His anguish over their pitiable circumstances and the epidemics that routinely rained over the communities planted in him a desire—and later an obsession—to do something toward the improvement of humanity. In order to wage such a war, you have to identify the enemy, and Reed found his: disease, filth and poverty.

One night he returned home to the apartment he shared with Christopher. Walter Reed stood for a long time at the dark window, his silhouette backlit by gas lamps. He thought his brother was sleeping as he stared out the black glass and said quietly: "Woe unto thee, Chorazin, woe unto thee, Bethsaida, for if the mighty works which were done in you had been done in Tyre and Sion, they would long ago have repented in sackcloth and ashes." The brothers never spoke of it; as Christopher later explained, "There are some things too sacred for the invasion of words."

On a trip to North Carolina where his family now lived, Reed met his wife, Emilie Lawrence, the daughter of a prominent planter. Emilie, who changed the spelling from *Emily* to be more elegant, was pretty, though not conventionally beautiful. She had a long, straight nose, honey-colored hair and lean lips. Reed wrote of Emilie's "beauty of character, your womanly worth, the purity of your Christianity, the charm of your intellect."

"Should you spurn my affection," he wrote to her, "I should never be ashamed of having revealed my love to so noble a woman."

It was upon meeting Emilie that Reed's work in New York lost its luster once and for all. He wrote to her that if he "could find some fair damsel, who was foolish enough to trust me, I think I would get—married, and settle down to sober work for the rest of my days in some small city where one could enjoy the advantages of a city and at the same time not feel as if lost."

Reed wanted to apply for a position as a surgeon in the U.S. Army. He lacked the wealth and connections for a viable medical practice on his own—medicine at that time had much more to do with social position than actual education. What's more, Reed's age was proving to be an obstacle. He wrote to Emilie: "It is a remarkable fact that a man's success during the first decade depends more upon his beard than his brains." Army medicine would offer stability and direction to the profession, and it would provide Reed with the opportunity to do research and to travel. "Who would have thought ten years ago that one of us would want to wear the army blue?" his brother Chris remarked.

Reed studied for months before appearing in front of the examining board of the Medical Corps—an exam that would take thirty hours over the course of six days. The exam would cover not only medical topics, but also history, mathematics, Greek, Latin and even Shakespeare. One question from the examination dealt with a timely topic—yellow fever. Reed answered in handwriting neater than usual that the disease was spread by poor sanitation and prevented by quarantine.

Out of 500 candidates, Reed was one of the 30 commissioned as an assistant surgeon. After a short stay on Long Island, Reed would report to the Arizona territory. This presented a new problem.

On the frontier it would be hard for Reed to have a wife, especially a wife who had been born and raised in the East. First Lieutenant Walter Reed traveled to Washington to meet with the surgeon general, "His Highness," as he called him in letters to Emilie.

"Sit down," the surgeon general barked at Reed. "What do you want?"

"I would like to know, General, if I can get a leave while I am in Arizona. I am engaged to be married and would want to return."

The surgeon general looked at the young soldier in front of

him and gruffly told him, "Young man, if you don't want to go to
Arizona, resign from the Service."

"General, I did not labor for my commission with the will I
did to throw it away so hastily, nor can you deprive me of it till I
act so unworthily as to cause dismissal."

The surgeon general's manner changed. "Have a cigar, Dr.
Reed, and let's talk it over. This is the advice I give you. Don't
marry now—go to Arizona and no doubt some soldier will become insane, you can bring him east, there will be your chance for
your wedding."

Reed decided against the surgeon general's advice and married Emilie on April 26, 1876, taking her west with him. It was a
wise decision; it would be another thirteen years before he was ordered back to Washington. The Reeds' assignment would not end
with Arizona: They served a short stint in Maryland; three different camps in Nebraska; and finally Alabama, where Reed looked
after several hundred Apache Indians, including Geronimo. During his years on frontier camps Reed would deliver both his son
and his daughter, and the family would adopt an Indian girl named
Susie.

As the family physician for frontier camps, Reed worked among
the soldiers and neighboring Indians binding wounds, performing
surgery, treating venereal disease and caring for soldiers and their
families when an epidemic coursed through the camp. Reed and
Emilie both loved to garden and planted what vegetables and fruit
would grow in the arid soil. In the Victorian era, gardening had become en vogue. In keeping with ornate Victorian architecture and
the fairly new emphasis on outdoor recreation, people planted
beautiful, well-kept gardens and green lawns. But Reed also saw
gardening as a health benefit, planting fruits to cure scurvy, herbs

for treatments and handing out vegetables to accompany the mostly game diets of the soldiers. He also strictly enforced cleanliness, believing it to be essential to preventing the spread of disease.

Nonetheless, medicine was in a period of great transition, and Walter Reed was missing it. In Europe, use of the microscope had given birth to the new field of pathology, finally separating general patient-focused medicine from real science. Laboratories had become the epicenter for experimental medicine. And medical education had turned from mere lectures to investigative study. The United States had decided to join, at last, the progressive movement in medicine.

Johns Hopkins had been established in Baltimore in 1876 as a school dedicated to modern science, modeled after European institutes, particularly the German ones. Germany had long been at the forefront for not only medicine, but their system of education. In Germany, knowledge was readily available, and students could move freely from one university to another, picking and choosing their subjects. British and American doctors flocked to Germany, and later France, for this independent, experimental system of study. In Europe, physicians and scientists could focus their energies on evidenced-based medicine and the life sciences in a culture unencumbered by theocratic shackles. It was time for America to offer its own institute for such study, but like many progressive ideas, it would be met with controversy.

When Thomas Huxley, the great evolutionist, was asked to speak at the launch of Johns Hopkins University, it outraged Baltimore society. One minister wrote: "It were better to have asked God to be there. It would have been absurd to ask them both." In his book, *The Great Influenza*, historian John M. Barry wrote that Huxley did not even mention God in his speech, and it came in an era when "American universities had nearly two hundred endowed chairs of theology and fewer than five in medicine." Barry added, "In no area did the United States lag behind the rest of the world so much as in its study of the life sciences and medicine."

Though the Johns Hopkins medical school would not officially open until 1893, its research labs were in full operation. Finally, America had a place for real scientific study, and in the following two decades, nearly every distinguished physician or scientist would come out of those labs.

The new American frontier extended beyond geography; it influenced medicine as well. Doctors began to push the limits of science beyond the boundaries of the Victorian era, into the Progressive one. The United States would finally have the opportunity to produce its own Pasteur, Lister, Koch.

Walter Reed had been out of touch with the world of medical research for nearly fifteen years. By 1890, when Reed, Emilie and their children returned to the East Coast, he ached to learn of the latest advancements, begin work in the lab and attend lectures. Reed was granted permission to begin studies under Dr. William Welch, a doctor only one year older than Reed, who had studied in Germany with Robert Koch and Paul Ehrlich. Balding and bewhiskered, Welch was known as "Popsy." One historian wrote: "No one symbolized the Germanic spirit in American experimental medicine better than Welch."

Welch had been handpicked by John Shaw Billings to join the staff at Johns Hopkins in 1884, and he is largely credited with selecting or influencing America's greatest medical minds—William Osler, William Halsted, Howard Kelly, Simon Flexner and Walter Reed. Under Welch, Johns Hopkins attracted brilliance the way a lens draws light.

Reed worked with Welch in the "Pathological," as they called their lab. Rather than listening to instruction, they engaged in learning, studying and identifying bacteria beneath the microscope, dissecting animals and experimenting. The Hopkins doctors would break from their work to lunch on wild duck or fish at the "church," a pet term for the local tavern that seemed more appropriate to scientists than a pub.

Reed thrived in the new atmosphere and appealed to a superior, John Shaw Billings, to stay at Hopkins. His request was denied, and he was sent, yet again, for a short stint at various forts, binding wounds from the Indian Wars and thwarting epidemics.

In 1893, a new surgeon general was selected: George M. Sternberg. Like Reed, Sternberg had been serving on the frontier, but more than simply treating disease, he was studying it, creating labs on dusty frontier outposts. Sternberg was not just an army man; he was a scientist. "The fossil age has passed," wrote Reed.

Sternberg opened a new Army Medical School in Washington, D.C. Reed was promoted to the rank of major and asked to join the faculty, as well as serve as the curator of the Army Medical Museum, a position recently vacated by John Shaw Billings. Reed and his assistant, James Carroll, taught bacteriology, while continuing their lab work, shuttling back and forth between Baltimore and Washington. Knowledge unfurled before them; new studies and papers appeared almost faster than they could keep up.

Reed had worked among the filth in America's largest city. He watched epidemics of infectious disease consume entire communities. He had labored in frontier camps and on army bases, observing the conditions of soldiers, their habits of hygiene. In his research, he quickly moved beyond simply identifying disease toward understanding ways to prevent it. Reed's work as both a practicing physician on the frontier and a scientist at the center of the latest theories made him uniquely qualified for investigative medicine.

Surgeon General George Sternberg's first commission for Walter Reed came during the Spanish-American War—war had a way of bringing science and soldiers together. Reed was to join Victor Vaughan and Edward Shakespeare in a study of typhoid. In spite of the fact that Sternberg was a staunch believer in sanitary practices, the army had been slow to act, and the surgeon general was now under heavy criticism for it. Thousands of soldiers crowded together in filthy camps waiting for orders; filth was expected in foreign camps, but it was scandalous that camps on American soil were in such poor shape. Ninety percent of the regiments reportedly contained cases of typhoid where the disease had been known to infect as many as one in four soldiers. "The government," one newspaper column read, "may well consider the propriety of ordering all the troops at that point to Cuba or Puerto Rico for the improvement of their health." When Reed's Typhoid Board arrived at Camp Alger in Virginia, they found that the hospital barracks had no microscope. No autopsies had been performed. The camp doctors routinely confused cases of typhoid for malaria. But worst of all, the camp was filthy.

When the commission visited another camp to inspect sanitary conditions, the colonel gave them a tour of the grounds, which were foul. Reed turned to the colonel and said, "Shakespeare and Vaughan are on this commission because they know something of camp sanitation. I am here because I can damn a colonel." Then, he lectured the officer on the responsibility an officer has to enforce good sanitation and health among his troops.

The worst camp they visited was Chickamauga Park on the border of Tennessee and Georgia; it was the largest camp in the country, and its name literally means "River of Death." When the commission first arrived, the better part of 60,000 troops had just moved out, and piles of human excrement littered the ground at every step, soaking into the soil and water supply.

The common element in all of the camps was poor sanitation. The commission studied the conditions at all of the camps, plotting the arrival and departure of new troops, the supply of water, the food prepared in the mess hall and the orderlies who would empty bedpans, then eat lunch without washing their hands. The board sprinkled lime on the latrines and then watched as the same flies landed on food in the mess hall, their fine-haired legs tinged with the disinfectant. Finally, the commission put all of the pieces together.

Typhoid, the commission showed, was not only infectious, but also contagious. It seeped into the water supply through poor sanitation and spread among the troops through poor hygiene. Departing regiments left scores of infected tents and bedding behind when they moved out, and a new regiment soon occupied them. And flies carried the bacteria all through the camp, depositing it in new places. The Typhoid Commission found the order of feces, filth, fingers and flies as the agents spreading the disease. With basic sanitation practices, the camps could improve their odds against typhoid dramatically.

With new sanitation measures in place, the surgeon general had great hopes for eliminating not only typhoid but other diseases as well—especially yellow fever. The Spanish-American War had ended, but troops remained, occupying Cuba and dying of disease. Though typhoid and dysentery had proven to be most disastrous, yellow fever was the most feared. The haunting stigma surrounding the American plague remained. Soldiers who had never even seen it feared it above all others; stories of the 1878 epidemic still circulated. "Medical men for centuries had been working in the dark, not knowing what to fight. That fact alone made yellow fever the most fearful of all diseases," wrote one army doctor.

Sternberg's hopes perished with the onset of another yellow fever epidemic in Havana in 1900.

Cleaning up the camps had done nothing to slow the spread of yellow fever. Two hundred years of scourges and two decades of Surgeon General Sternberg's personal study had yielded no definitive answers about the disease. So Sternberg handpicked a frontier physician who had shown remarkable promise in the lab and swift success with typhoid. Sternberg had a special assignment in mind for Reed, and he would soon report to a camp just outside of Havana called Camp Columbia.

Victor Vaughan and Edward Shakespeare would be left on their own to finish the typhoid report. Vaughan would hardly fade behind the fame of Walter Reed. His career had only begun— Victor Vaughan became one of the heroes in the fight against the 1918 influenza pandemic, and one historian would call him "second only to Welch in his influence on American medical education." Still, Vaughan would also learn a valuable lesson, one that many other doctors failed to realize. "Never again allow me to say that medical science is on the verge of conquering disease," he later said.

CHAPTER 12

A Meeting of Minds

Everything about Camp Columbia was vivid. The colors were bold: bright blue sky, yellow heat, the flat swath of green, the clapboard buildings standing row after row like game pieces carved from bone. Though several miles from shore, the white light of the camp was suffused with salt and sea wind. The scent of over-ripened pineapples in nearby fields sugared the air, but beneath the sweetness, the chemicals were pungent. The smell of disinfectants, iodine and alcohol crossed the breeze. The landscape appeared vast, unbroken, except for the shadeless royal palms with their gray trunks standing like granite columns among the buildings. At the northeastern point of the camp, the Gulf of Mexico stretched in staggered shades of blue. The weather existed in extremes as well with only two seasons: wet or dry.

Camp Columbia's focus was the care and treatment of sick soldiers. Three hospital buildings stood on stilts of varying heights that leveled the walking porches connecting them, and walkways of white limestone led to the main entrance. One housed a convalescent ward. Another held a small restroom and bathing rooms connected by a large iron pipe to funnel in heated water. There was also a surgical ward with eighteen beds, an operating room and dressing rooms. And one ward, lined with thirty beds and fine mesh netting, held the typhoid and undiagnosed fever patients. Several hundred feet away, across the La Playa road and a footbridge, even past the railroad tracks on the outer edge of the Columbia Barracks, were eight small huts. They were the yellow fever wards.

The camp had been built on a high plain about six miles west of Havana near the town of Quemados, Marianao, and it housed 1,900 American soldiers—part of the 15,000 men and women left behind to occupy Cuba following the war. The men woke early to the bugle calls, working in the hospital wards until noon. After lunch and an hour of "bunk fatigue," the doctors, nurses and enlisted men were back at work.

The enlisted men came from all parts of the United States. They were young, many only twenty or twenty-one years old, and made to look even younger by their clean-shaven faces. With the advent of disposable razor blades, the heavy beards of the last century were gone, and facial hair now signaled status. A moustache was an indication of rank in the army. The bigger the moustache, like Teddy Roosevelt's walrus whiskers, the higher the rank. Most of the officers went for something in between, opting for a pencil moustache or handlebar style. And accordingly, the enlisted men went clean shaven.

For entertainment, the men traveled into Havana, where they saw ball games and bought pastries and snacks like Bermuda onions and Norwegian sardines. Sometimes, they bought brandy.

The Cubans, who were accustomed to the local cocktails, found the drunken, staggering Americans very amusing. There was also the occasional afternoon swim off La Playa one mile away from Camp Columbia. In the evenings, classical music from the camp's band filled the night sky.

Dr. Albert Truby first arrived at Camp Columbia in 1898, sailing into the Havana harbor, where Spanish soldiers stood along the Cabaña Fortress, and Cuban locals sat along the seawall to see the American ships arrive. It was December, and Truby and his men wore their winter wool uniforms, now damp with sea air and sweat. Under his watch, Truby had 1,000 soldiers—all nonimmunes.

Truby looked like a doctor. He wore small, wire-rimmed spectacles on a round, almost cherubic face. His hairline was receding, and in general, he appeared gentle. Truby had just earned his commission from the medical examination board at the Army Medical Museum in Washington, D.C. As Truby and the applicants stood before the board, he noticed a tall, slender officer. The officer was especially interested in the applicants' knowledge of malaria. Prepared slides were placed beneath the microscope, and the applicants were asked to identify common bacteria and malarial parasites. The officer was quiet and courteous. He seemed like a teacher, one who knew his subject very well, and he gave Truby a feeling of confidence. When asked about malaria, Truby remarked on the recent work proving that mosquitoes can transmit the fever. The officer seemed momentarily pleased then continued the questioning. Only two of the applicants were accepted that day, and Albert Truby was one of them. He would be recommended to the surgeon general for an appointment in Cuba.

Truby thanked the medical examination board and the slender officer who had been so interested in malaria. It was the first time Albert Truby met Walter Reed. The second time they met was in Cuba.

Although he had never seen a case of yellow fever prior to his appointment, it was Truby's job to field all fever cases. The patients, taken to the mesh-wire receiving ward, had their blood screened for malarial parasites and their urine tested for albumen. There were local doctors, self-described yellow fever experts, but they rarely showed the talent to back up the claim. There were two doctors, however, that Truby and other contract surgeons relied upon for a yellow fever diagnosis. One was Carlos Finlay, and the other was Juan Guitéras; both had been members of the original Yellow Fever Commission in 1879.

The surgeon general sent Walter Reed to Cuba in March of 1900. Reed left the typhoid report, stalled due to a lack of funds, in the capable hands of Victor Vaughan and Edward Shakespeare. Prone to bad bouts of seasickness, Reed, on the advice of a friend, took a dose of bromide before they set sail for Havana. He retired to his stateroom and went to sleep. When he awoke, he was impressed with the medication. He didn't feel sick at all, until he went up to the deck and realized that the ship had not yet set sail. For the next two days, he spent most of the voyage sick in his cabin.

Reed was sent to the Columbia Barracks to investigate a new disinfectant made from seawater called electrozone. Tank wagons rolled through Havana showering city streets with the expensive, salt-laced sanitizer. The disinfectant didn't impress Reed, but Camp Columbia and its physicians did.

Reed stayed in a new bungalow flanked by a wide veranda on

each side. The building served as quarters for bachelor medical officers. Its clapboard frame shored up a large tile rooftop that draped over the edges of the wide porches where rain would soon fall in fixed streams. Shutters on every wall allowed for a breezeway, and in some of the nicer buildings the windows were made of glass. In many, however, there were only the wooden shutters to close against the rain. It was on this veranda that Reed and the other doctors stationed at Camp Columbia sat in their army whites and listened as the band's music drifted in the approaching dusk. The insects were not yet prolific, but soon, as the wet season approached, the hum of mosquitoes would hang on the melody of the nighttime music.

Albert Truby stood on the porch with Reed, as did two other contract surgeons. Reed's good friend Jefferson Randolph Kean, chief surgeon of western Cuba, was also there, as was Dr. Aristides Agramonte, an American doctor of Cuban descent. Agramonte had worked with Reed in Washington before being sent to Havana.

Another doctor sat on the veranda that night. Though he looked younger than the other men, he was an assistant surgeon for the army and the resident bacteriologist who headed the camp's lab. Just a few months before, Albert Truby had put in a request to Major Kean for a full-time bacteriologist to head the lab at Camp Columbia, and the army appointed Jesse Lazear.

Dr. Jesse Lazear was from the East, born and raised near Baltimore at Windsor, the estate of his grandfather. He had a close-cropped beard and black hair offset by blue eyes. He rarely spoke and asked questions even less. Lazear was well liked by everyone, even described as lovable by a number of the men. Agramonte, who had been a fellow classmate at Columbia, would later write of Lazear, "A thorough university man, he was the type of old southern gentleman, kind, affectionate, dignified, with a high sense of honor, a staunch friend and a faithful soldier." The word most often used in reference to him was *gentleman*, and in that age of moral high-mindedness, manners and codes of conduct, it was the greatest compliment one could bestow. Jesse Lazear was the type of man who wrote to his mother every day and loved Cuba because he didn't have to play golf.

At only thirty-four years of age, he was highly accomplished, and often, more qualified than medical officers who outranked him. Lazear attended Washington and Jefferson College, graduated from medical school at the College of Physicians and Surgeons at Columbia University in New York, as well as Johns Hopkins University. Soon after graduating, he left for Europe to study with the greatest minds in science at the Pasteur Institute. His studies continued in Scotland, Germany and France, where he wanted to improve his language skills. As most groundbreaking scientific studies came out of Europe, they were usually in German or French, so Lazear wanted to read and understand them in their original language. When he sailed back to the States, he returned to Baltimore, where at the age of thirty, he became the first doctor in charge of a clinical lab at Johns Hopkins.

While Lazear was a physician who would be talented in any number of fields, lab work was particularly suited to his ability. He was gifted with perception, a sort of insight for the way things work, and he was meticulous. His thick glass slides, smeared with blood, were kept in perfect order in a wooden box alongside a leather logbook for notations. Next to the slides sat his microscope, which resembled a spyglass. It was his attention to detail, his obsession with accuracy, which would prove to be the haunting mystery left in Cuba long after the doctors departed the island.

* * *

As Lazear sat on the veranda that warm March night, he and the other men watched Reed with a sort of reverence. Not only Reed's rank but his reputation impressed colleagues. He was twenty years Lazear's senior, and a close friend of Welch and the surgeon general. Lazear also felt a sort of kinship in their work; they had worked in the camp lab often. Walter Reed even visited Jesse Lazear and his wife at their home in Cuba. "Mabel," Lazear proclaimed, "I have another convert. Major Reed also believes the mosquito theory."

Lazear was enthusiastic about his work with yellow fever and the theory of Cuban physician Carlos Finlay, the scientist who had put forth a theory that the fever was spread by mosquitoes. Once the connection between malaria and mosquitoes was made, Finlay's theory seemed all the more plausible to Lazear. Finlay was thrilled when Lazear approached him about his theory. Finlay was now in his sixties with long burnside whiskers. It had been twenty years since he first proposed his theory; the mad scientist would finally be given the chance to be taken seriously.

As the sky grew plum colored, and the music had long since silenced, Reed sat with the other doctors on the veranda and talked about medicine. But most of all, they discussed yellow fever. Reed's interest was tireless.

When Reed finished his investigation of electrozone, he sailed back to the States to give his opinion of the disinfectant to Surgeon General Sternberg. Though Sternberg had sent Reed to Camp Columbia on orders to examine the disinfectant, it's more likely that he wanted to pique Reed's interest in yellow fever. It is not known whether Reed approached the surgeon general about a yellow fever study, or if Sternberg proposed the idea to Reed, but in the end, it was a moot point. In less than two months time, Reed would be on his way back to Camp Columbia.

After Reed left Cuba, Lazear's work continued as it had before. He spent his days in the hospital or lab and his nights at home with his wife, Mabel, and their one-year-old son, Houston.

Jesse Lazear met Mabel in Europe, where they were both traveling with their mothers. Mabel was described as a young lady with expressive eyes and an intriguing introspective appearance. Like Lazear, she had a love of the outdoors, and at her family ranch in California, Lazear and Mabel had enjoyed trout fishing, hunting and climbing together. They married on September 8, 1896, in San Francisco when Lazear was thirty and Mabel was twenty-two years old. They settled in Baltimore, and Houston was born three years later. All three moved to Cuba the following year when Lazear joined the army as a contract surgeon. Considering the Victorian age, it is not unusual that there is no mention in any correspondence about the fact that Mabel was four months pregnant with their second child when they arrived in Havana.

Lazear thrived during those first few months in Cuba. He wrote to his mother, "We were surprised to find Havana a most beautiful city, entirely unlike anything we had ever seen before. The color effects are charming—wonderful greens and pinks. There are numerous fine gardens with magnificent palms and flowers."

Lazear kept a ribbon-bound photo album of their stay in Cuba. The recent invention of the Kodak box camera allowed for the first time average people to take snapshots of their lives candidly, rather than posed formally in front of a photographer. In his album, Lazear pasted pictures of Cuban landscape, Havana street scenes and his family: Houston toddling through a grass field; Mabel, with the sea air blowing strands of her hair from its bun, holding Houston and a toy guitar on the beach. The last photo in the album is a loose-leaf one of Jesse Lazear. He sits on a wooden fence with mountains in the background. The snapshot is a striking difference from the posed, stiff photos of recent times. His posture is relaxed as he perches on the top board of the fence, his two feet balancing him. His arms rest on his legs, his hands folded one over the other in the middle. He is dressed casually with a cap tilted on his head while he smokes his pipe. There is a hint of a smile on his face. He looks content, happy even.

Lazear's living quarters were not quite what they were in Baltimore. He quickly wrote to his mother to unpack the boxes of golf clubs, books of Shakespeare, linen napkins and dishes. They need not be shipped. Much in the style of the other buildings at Camp Columbia, Lazear's family quarters consisted of a two-room house made of wide, pine boards. A wooden bridge connected the two wings. In one house, Lazear and his wife lived; in the other, Houston and his nanny, Gertrude, slept. There was a roof and rafters, but no ceiling. There were shutters, but no glass windowpanes. A sloping roofline covered the walk-around porch.

The small shower bath had its own dwelling about one hundred feet away from the main house, but Lazear had hopes of having a real bathroom affixed to the house that autumn. The shower bath was in essence a bathtub with sprays of water that came out the side, so that one could shower without wetting one's hair, as one soldier described it.

Mabel had done her best to make it feel like a home. Shopping in Havana, she had found matting at a Chinese store to sew and hang as a partition, giving them a bedroom on one side and sitting room on the other. Mosquito nets, while practical, also added a web of gauze to the otherwise hard, plank-wood bedrooms. Even with the netting, nature could not be contained. Fleas would often bite the baby. And tree frogs settled into the rafters, falling with a damp thud against the beds, sometimes landing in the water bucket. Soldiers would often begin their morning shave only to look down and find a tree frog with all four feet sucking the side of the pail and its head barely above the surface of the water. Eventually, covers were issued for the water pails.

The surrounding countryside was the real charm of their situation. Only a few miles from the beach, Lazear went, almost daily, for a sea bath. Sea grapes and mangroves tangled the shoreline, where white sand sloped toward a green-blue ocean. Houston played in the sand and collected shells. Every afternoon, Gertrude took Houston on a long walk in the countryside and let him chase chickens.

Carts of fresh produce or mules strapped with baskets of vegetables regularly came into the camp from Havana. Fresh meats were shipped from Chicago, packed in ice. Mabel had brought Borden's condensed milk for the baby. Houston also had a healthy supply of oatmeal, eggs and meat juice.

As the rainy season, and more important, the quarantine season, approached, Mabel and Houston planned to sail for the U.S. In her sixth month of pregnancy, her condition was certainly a factor, but there was also a more practical reason. Once quarantine was under way, the fumigation process in New York would ruin all of Mabel's clothes and personal belongings. On April 14, Lazear took Mabel and Houston to the Havana harbor to ship out on the steamer *Sedgwick,* where he bought her a twelve-dollar ticket and said good-bye. Jesse Lazear probably had another reason to send his wife and son away—locals were already referring to this one as a yellow fever year. It must have been a sad parting. The fever season would last several months, and Lazear's work in Cuba would keep him too busy to travel back anytime soon. Houston would grow and change during those months away, and most likely, Mabel would give birth to their next child before the family could be together again.

Lazear continued with his daily work in the hospital wards and lab after Mabel and Houston left. He swam in the sea and ate with the other officers in the mess hall, where they drank red wine in an attempt to keep fever at bay. The men entertained themselves with cards, a brass spittoon at the foot of each chair, or on special occasions, smoked an old Madre rolled cigar. Potted ferns and palms climbed the walls of the social hall, as though the flora of Cuba would not be kept out. Open shutters and high ceilings crisscrossed in wooden beams helped keep the room cool. Dances were held there on Saturday nights, and as always, the music continued to infuse the tropical night. Lazear listened from his porch in the dark, though he rarely walked to the dance hall unless it was a clear, moonlit night. Without a full moon, the tropical dark felt oppressive with only the patchwork of yellow window light and pinpoints of starlight to break up the blackness. And on quiet nights, he could hear the sounding of the hour fired from El Morro Castle.

On May 1, Major Jefferson Randolph Kean began keeping a journal to record any cases of yellow fever. Kean was the chief medical officer for western Cuba, and he lived in Quemados, Marianao. Kean was also a close friend of Walter Reed's. Both graduates of the University of Virginia, the two met in Key West investigating a smallpox outbreak. Kean found Reed's "whimsical humor" and penchant for "quaint stories" entertaining, and they would become lifelong friends. The two had even exchanged frustrated letters when Surgeon General Sternberg had denied their placements in Cuba when the Spanish-American War broke out. Sternberg did not want to risk two of his best medical officers; neither had ever had yellow fever.

On May 21, Kean recorded in his diary that two cases of yellow fever appeared on General Lee Street, several blocks apart and in homes that had no contact with one another. General Lee Street ran through Quemados, a town of rainbow-colored houses set against ripe hillsides and thickets of tropical plant life. Palm fronds and bougainvillea blossoms, like fuchsia petals of parchment, enclosed the homes, one of which held the feverish wife of a cavalryman. She was too afraid to call the doctor, even as the bleeding began, for fear of being sent to die in the yellow fever ward.

Two days later, Lazear was called to No. 20 General Lee Street to investigate Sergeant Sherwood. When Lazear arrived, Sherwood was running a temperature of 100.4 and complained of a headache. By the next day, Sherwood's temperature rose to 102. Lazear suspected the worst, quarantined the house and sent the sergeant to the yellow fever hospital. He conducted the Widal test to rule out typhoid and studied the blood for malarial parasites, but Sherwood's skin grew mustard colored and his gums began to bleed. By nightfall, he was delirious and slipped into a coma, his breathing heavy and strained. The following day, May 30, Sergeant Sherwood died at 11:30 a.m. Lazear autopsied the dead soldier, making comments in his notebook: "Extreme jaundice, peculiar mucus like applesauce, liver was a bright yellow color, stomach contained about a pint of black coffee ground fluid." It was a clear case of yellow fever. Another twenty-three cases would quickly arise in the town of Quemados.

Lazear also kept detailed records of the mosquitoes beginning to swarm in May, sending samples to an entomologist in the United States. Lazear's meticulous nature was perfectly suited to this sort of study; as described by one tropical medicine professor: "keying in an identity depends on anatomical minutiae—how the insect's hairs are placed and grouped, the formation of the mouth parts, the sex parts, the bewildering pattern of wing venation." During this time, Lazear began killing and dissecting his pet collection of mosquitoes, or "birds" as they were nicknamed, most of which, he noted, had striped legs and bodies.

The fever appeared dangerously agile, jumping from one house to another, traveling from Calzada Real and back to General Lee Street. On Real Street, in close proximity to No. 20, a saloon and a number of local bordellos were shut down when it looked as if the risk of yellow fever was greater in those men who frequented them. For physicians trying to track the disease, it proved evasive and unpredictable, as if it engaged their interest as sport. "This epidemic," wrote Truby, "with fifty cases and twelve deaths in one of the finest and most sanitary villages in Cuba disturbed everyone and left a lasting impression."

On June 21, 1900, the entries in Jefferson Kean's diary came to an abrupt stop.

A few days earlier, Kean had learned that a friend and neighbor was down with yellow fever. Kean had been ordered to stay out of the infected district, but early one morning, he decided to make a visit to his sick friend. He took every precaution, never entering the infected house, and instead sat outside on the porch where the air was clear. "I obeyed the letter but not the spirit of the order," Kean would later write. He spoke to a nurse through the iron bars of the open window. He never came in contact with any of the infected items, nor with his friend. Kean was shocked, five days later, when he fell feverish and was admitted to hut number 118 in the yellow fever ward of Camp Columbia.

CHAPTER 13

The Yellow Fever Commission

WAR DEPARTMENT,
Surgeon General's Office

Washington, May 23, 1900.

To the
ADJUTANT GENERAL OF THE ARMY.

Sir:

I have the honor to recommend that Major Walter Reed, Surgeon, U.S. Army, and Contract Surgeon James Carroll, U.S. Army, be ordered to proceed from this city to Camp Columbia, Cuba, reporting their arrival and instructions to the commanding officer of the post, the commanding general, Department of Havana and Pinar del Rio and the commanding general Division of Cuba.

I also recommend the organization of a medical board, with head-

quarters at Camp Columbia, for the purpose of pursuing scientific investigations with reference to the infectious diseases prevalent on the island of Cuba. and especially yellow fever. —Stricken

The board to be constituted as follows:—Major Walter Reed, Surgeon, U.S. Army; Contract Surgeon James Carroll, U.S. Army; Contract Surgeon Aristides Agramonte, U.S. Army; and Contract Surgeon Jesse W. Lazear, U.S. Army.

Contract Surgeon Agramonte is now on duty in the City of Havana and Contract Surgeon Lazear at Camp Columbia. It is not considered necessary to relieve them from the duties to which they are at present assigned.

The board should act under general instructions which will be communicated to Major Reed by the Surgeon General of the Army.

Very respectfully,
George M. Sternberg,
Surgeon General, U.S. Army

On the evening of June 25, 1900, Walter Reed sat on the deck of the *Sedgwick* and wrote a letter to his wife. A chill imbued the inky sky, and Reed fastened the overcoat his wife had sent aboard just moments before the steamer set sail from New York that morning. He had not thought to pack it; after all, he would hardly need it once he arrived in Cuba.

The unfinished letter to his wife would take several more attempts to finish, which Reed literally chronicled as *Effort no. 1, Effort no. 2, Effort no. 3* at the tops of the pages. As with most of Reed's voyages, he would spend much of it sucking lemons and eating crushed ice to keep the motion sickness at bay. Regardless of his efforts, and regular doses of bromide, Reed consistently "fed the fish" over the railing of the boat, losing five pounds on the voyage.

As the breeze began to warm and clouds gathered over the green fringe of the Florida coast, Reed managed to keep down an orange, a cup of coffee and dry toast. A ribbon of rain showers lined the coast, and from the deck, the men watched schools of flying fish and porpoises chase the *Sedgwick* as the 5,000-ton steamer barreled toward Cuba.

In the wan morning light, before the sun had burned off the haze, the buildings of Havana appeared in shades of gray and blue, wedged between the dark sea and pale sky. But as the light rose, the buildings brightened, and the weary stone of El Morro Castillo warmed, incandescent bursts of green growing amid its stones. Waves knocked against the fortress to one side of the harbor and against the seawall on the other as though the sea itself were sleeping, its breast rising and falling in heavy, rhythmic breaths.

Reed sat on the deck, again writing a letter to his wife, and watched Havana come into focus, smelling the salt, steam and wet stone, and farther off, the scent of smoke, coffee and old hay. The harbor blazed with color: The flags of nations all over the world whipped in the breeze, white sails skimmed between steamers, and green treetops glowed against cobalt-colored mountains far in the distance. Then his eyes fell on an iron corpse, mostly submerged but for a tangle of beams like splintered bone, wires flailing and an American flag at half-mast. His handwriting grew wilder and slanted as he wrote, "The City of Havana from the shore is certainly very beautiful, but as I write I see the wreck of the *Maine* not more than 400 yards away, and it makes my very blood to boil—The whole Island wasn't worth the loss of those brave men & gallant ship—Damn every Spaniard that ever lived!"

* * *

On May 21, the same day that Jefferson Kean had recorded the first two cases of yellow fever in Quemados, Surgeon General Sternberg had put in a request in Washington, D.C., to form a board to examine yellow fever. Though the directive would be to study "all infectious disease" afflicting the camps in Cuba, Sternberg made sure, verbally, that the focus would be yellow fever.

Sternberg had assigned Walter Reed and James Carroll to probe the finding of Dr. Sanarelli, an Italian bacteriologist who claimed to have found the microbe that caused yellow fever. For almost a decade after the congressional committee's conclusion that bacteria caused yellow fever, medical theories and experiments on the subject of the disease stagnated until, on July 3, 1897, the *British Medical Journal* published an article about Dr. Giuseppe Sanarelli's discovery of *Bacillus icteroides:* the bacteria that caused yellow fever. Sternberg's pride had been wounded. Sternberg had missed his opportunity for fame with diseases like tuberculosis, pneumonia and malaria. A skilled microbe hunter, he had a personal passion to find and solve the yellow fever question, and after Sanarelli made some caustic remarks about a germ Sternberg had discovered, it became a battle of the egos. It also became another clash between the U.S. Army Medical Corps and the Marine Hospital Service. Sternberg was surgeon general of the Army Medical Corps while the Marine Hospital Service backed Sanarelli. The confrontation would continue for years.

A fresh outbreak of yellow fever in Havana and nearby camps provided Sternberg with the opportunity to test Sanarelli's new bacteria. He asked Reed to investigate it, and soon thereafter, to head the Yellow Fever Board. Though Sternberg would later take credit for recommending the three other members of the board, it seems more likely that Reed himself chose or at least suggested them. After his return from Cuba in April, Reed had submitted his report on electrozone to Surgeon General Sternberg, ending with, "In carrying out the experimental part of this report, I desire to state that I have received valuable aid from Acting Assistant Surgeons A. Agramonte, Jesse Lazear, and James Carroll, U.S. Army." Appointment to the board would forever change the lives of the four doctors.

First choice and second in command was James Carroll, Reed's longtime assistant. Carroll was by far the most eccentric of the group. The men called him "Sunny Jim" because of his bald head. Born in England in 1854 to a working-class family, his background was a hodgepodge of professions. Originally, he planned to enter the British Army as an engineering student; instead, he was a self-described "wandering good-for-nothing who fell in love at fourteen and left home at fifteen, roughed it in the Canadian backwoods for several years and finally drifted into the Army."

After his love affair at age fourteen ended in heartbreak, he abandoned his plans for army life and emigrated to Canada where he worked at one time or another as a blacksmith's helper, a railroad laborer and a cordwood chopper. In 1874, he moved to the U.S. and joined the army as an enlisted man, a distinction that would color the remainder of his career and his ego. He would serve twenty-four years before wearing the uniform of an officer.

It was in the army that Carroll decided to pursue medicine. His was not a conventional background for medical school—he had no advanced degrees, he had taken one year of French and two years of German at a time when most doctors were fluent in a number of languages. Carroll began his studies in medicine as an apprentice to a doctor at Fort Custer, Montana. Carroll then applied to medical classes in New York and was at first rejected, then later allowed to attend classes at the University of the City of New York and the University of Maryland, finally graduating in medicine from the latter in 1891. He also took courses in bacteriology and pathology at Johns Hopkins, which placed him at the Army Medical Museum in Washington working as Walter Reed's assistant. At first, Carroll's fortune in achieving his degree and finding placement with such a highly respected physician pleased him. Later, it would turn into a lifelong torment.

James Carroll exuded an almost bitter work ethic. Jesse Lazear, in a letter to his wife, described him: "Dr. Carroll is not a very entertaining person. He is a bacteriologist pure and simple. To me bacteriology is interesting only in its relation to medicine. He is interested in germs for their own sake, and has a very narrow horizon . . . Carroll would amuse you very much. He is very tall and thin. Wears spectacles, bald headed, has a light red mustache, projecting ears and a rather dull expression."

Some colleagues found him reticent, crude and surly, while others described him as reliable and straightforward, though prone to "improper" behavior. He was kind and helpful in the laboratory, but also capable of profanity that would be the envy of any sailor, as one student put it.

Carroll proved in many ways to be the exact opposite of Reed, though their working relationship was described as warm and effective. Carroll's skill in the laboratory was undisputed, but he was a self-made man, intent upon self-improvement, who worked hard to achieve what men like Reed, Agramonte and Lazear seemed to accomplish with ease. Another colleague wrote, "Carroll was a most efficient worker, but he had to be led by a man with vision, like Reed."

Carroll's personal relationships also seemed in dramatic contrast to those of his fellow board members. Reed sent letters to his wife showered with terms of endearment and unabashed affection. Lazear was also a doting husband and an even more doting son. Carroll's letters to his wife, however, were cold and sometimes even cruel. In one, he chastised her for sending him fresh peaches when the price was probably exorbitant, and they would spoil anyway. In another he wrote, "Don't bother to send me any more letters that do not interest me."

In all, the image of James Carroll is that of a tragically conflicted man—a working-class Englishman in America; a soldier who was an enlisted officer, rather than a commissioned one; a doctor who was self-made; a frustrated husband who spoke of a lost love for the rest of his life; and a colleague who was innately proud to see his mentor recognized, while at the same time, riddled with envy.

It was a rancor that would extend to the next generation. After Carroll's death, his son hoarded the personal letters of his father, and many records belonging to Walter Reed, in disheveled trunks in his attic. When historians in the 1940s and 1950s attempted to acquire the material, he appeared paranoid and angry, refusing to accommodate a government that he believed had robbed his father of so much.

For the next member of his team, Reed chose Dr. Aristides Agramonte, a Cuban-born doctor working in Havana. Agramonte was young, only thirty-two years old. He looked the part of a Cuban gentleman with a waxed pencil moustache, pompadour hairstyle and aristocratic features. Some of the soldiers found him to be vain, but very capable. The two men had met before in the Hopkins labs and again during Reed's visit to Camp Columbia. Agramonte worked with Reed and Carroll for several months before the surgeon general ordered him to Havana to look for Sanarelli's bacteria in the autopsies of yellow fever patients. Born in Cuba, Agramonte was believed to be immune to yellow fever like most other locals who had been exposed to the fever in childhood. He had come to America when he was only three years old, after his father was killed in the first Cuban war for independence. His education in America continued, and he studied medicine at Columbia in New York, where he was a classmate of Jesse Lazear's.

Finally, Reed asked Sternberg to appoint Dr. Jesse Lazear to the board. He seemed like a fine doctor and a likable person. His work with malaria and yellow fever had impressed Reed. He also came highly recommended by Welch and William Osler, and there could hardly be any higher endorsement. William Welch, Lazear's mentor at Johns Hopkins, had written a glowing review of Lazear, describing him as "a good clinical man, a bacteriologist and withal a gentleman of enclivation and agreeable personality . . . and I am convinced that he would prove a very valuable addition to your corps of army surgeons."

Each of the four men could appreciate the skills of the other and respected one another. As individuals, they probably could not have achieved anything so great. Even Reed, the team's leader, had lacked the foresight to see a connection between mosquitoes and malaria early on. At a medical conference—on April Fool's Day—Reed refuted the idea that an insect could transmit disease. But what Reed lacked in vision, he made up for in an earnest devotion to knowledge and the scientific process. If there was a visionary in the group, it was certainly Jesse Lazear. With their talents pooled together, they knew there was a good chance of conquering this disease.

Michael Oldstone, in his book *Viruses, Plagues, and History,* wrote: "The obliteration of diseases that impinge on our health is a regal yardstick of civilization's success, and those (scientists) who accomplish that task will be among the true navigators of a brave new world."

With his team assembled, Reed was ready to begin work at Camp Columbia. He and Carroll left New York on the *Sedgwick* bound for Havana harbor on June 21, 1900, the same day Kean was stricken with the fever. His would be Reed's very first case of yellow fever. Walter Reed, head of the Yellow Fever Board, had never before seen a case of yellow jack.

Reed and Carroll landed on the busy Havana wharf where women carried metal pots of coffee and soldiers hauled supplies onto the shore, loading mule-drawn wagons. American voices mingled with foreign accents. There were shouted orders and laughter, neighing horses and barking dogs. Gulls called overhead, and buzzards flew high over the dirt roads leading out of the town, their black wingspans like line engravings in the blue sky.

Reed and Carroll rode by carriage into Havana, through Plaza de Armas where soldier tents surrounded the fountain and an American flag flew from the top of the Governor's Palace, General Wood's new headquarters. The carriage bounced along cobblestones, spraying dust as they passed Spanish churches and colorful, narrow town houses. Reed was excited to be returning to Cuba. Havana was unlike any place he'd seen before—the bones of the buildings unmistakably grand and European, while the fleshy surroundings were tropical and Caribbean. It was at one time elegant, and at another, raw. They passed palm trees, almond trees and gnarled jaguey. Plumes of red sprouted on poinciana. Hibiscus blooms erupted.

The road rose before them as they approached the elevated

outskirts of Havana and rode into Quemados, Marianao. The air seemed to thin as they moved farther away from the harbor's output of humidity, steam and smoke into lush jungle, and finally, the windswept farmland where Camp Columbia stood.

Reed went immediately to the hospital tents to see Kean. Kean's temperature was high, and his gums bleeding, but it would prove to be a moderate case, and when he was well enough, Jefferson Kean was shipped to family in New Jersey to recuperate.

On the evening of Reed's arrival, the Yellow Fever Board met on the veranda of the officer's quarters at Camp Columbia for the first time. It was a familiar scene for Reed, Lazear and Agramonte, who had stood there just three months before, only on this occasion, it felt much more formal. The officers, dressed in their army whites, listened as Reed quickly relayed the surgeon general's orders, which included the investigation of malaria, leprosy and unclassified febrile conditions. Reed stood straight-backed as he spoke. Whether speaking to a classroom of students or a handful of officers, Reed had the rare ability to be simultaneously amicable and commanding. A former student described Reed as someone who "spoke with great clarity, precision and force. His lectures were models in English and his scientific data were assembled in perfect order. There was nothing of the dramatic in his speech or manner, but he impressed one with his earnestness, thoroughness, and mastery of his subject."

The members of the board listened to Reed give the directive from the surgeon general with as much reverence and attention as his former students. Lazear had written to his mother that he was excited to be working with such an impressive and well-known man as Dr. Walter Reed. As the commission listened to Reed, another thought must have entered their minds: The crowning glory of this task would be yellow fever. To solve the American plague would be the greatest achievement of all. They agreed unanimously that whatever their discovery should be, it would be considered the work of the board as a body and not an individual achievement.

Reed instructed Cuban-born Agramonte, the group's pathologist and the only member thought to be immune to the fever, to stay in Havana in charge of the lab at the Military Hospital, where there would be no shortage of yellow fever autopsies. Agramonte had already done some impressive work on the Sanarelli bacteria, so he would continue that as well. Carroll, Lazear and Reed would remain at the barracks hospital at Camp Columbia, where Carroll would make cultures from various tissues, and Lazear would study them under the microscope. Lazear could continue with his mosquito investigations as well.

It is not known how seriously Reed considered the mosquito theory at this point. His letters to Emilie do not mention it, and he kept no records of the inner workings of the board. As an investigative scientist, he surely felt curiosity at the very least. Albert Truby believed Reed had been interested in the insect theory from the start, that Reed knew this was the time and the place to finally conquer yellow fever. But Reed was also a soldier, one who took orders, followed them and led men accordingly. That left little room for thinking outside the boundaries of a given task. An officer to duty first and foremost—and under pressure from Sternberg—Reed knew that the first order of the board should be to disprove the Sanarelli bacteria once and for all, putting that ongoing and very public controversy to rest. There wasn't time to deviate from the given course in search of a vector.

Just before he left for Cuba, Reed met with George Sternberg and discussed a possible connection between yellow fever and mosquitoes. "No," Sternberg answered. "That has already been decided to be a useless investigation."

CHAPTER 14

Insects

By the time the board had assembled at Camp Columbia, the yellow fever epidemic in Quemados was waning, if not entirely over. Yellow fever, it seemed, had taken its last victim for the season—though it left a lasting impression. Quemados, high on a plateau, had at one time been a safe haven for those fleeing Havana's fever epidemics. That year, it struck the beautiful suburb with ferocity.

The momentous beginning and the enthusiasm the Yellow Fever Board felt soon stalled. Without active cases to investigate, Reed focused on setting up a lab. Built in what had once been the camp's operating room, the lab became the focal point of their work. Along one wood-frame wall stood a tall table pushed against the window. Strong northern light fell in planes onto the microscopes of Walter Reed and James Carroll, who worked side by side. Against the far wall of the lab was another table covered in jars and lab equipment, like a glass menagerie of winged insects.

Off of the lab was a room for Reed's lab assistant John Neate, which contained shelves with rows of test tubes, jars of black vomit and an incubator. And across from it, another small room held two caged monkeys, supplies and guinea pigs.

On a wooden table in the center of the board's lab, Lazear kept his own pet mosquitoes in jars with sweetened water and bits of banana. He collected his "birds" from different areas: Havana, Pinar del Rio, the Post Hospital. But all had fed from the blood of yellow fever patients. And, for each mosquito, Lazear scribbled an entry into his red notebook, his handwriting becoming quicker, sloppier and abbreviated as he went.

It also took a little time for Reed to adjust to life in Cuba. Within days of arriving, he was already sunburned, and his northeastern clothes were too warm. Reed traveled into Havana to buy two linen crash suits and a cork helmet. It was hot, but not as hot as a Washington summer, he wrote to Emilie. That first week, Reed awoke at 5:30 or 6:00 every morning, sitting at the desk in his pajamas to write a letter to his wife. He always addressed the envelope the night before so he could drop the letter in the mail as soon as he heard the mail wagon. After a shower-bath and breakfast, Reed headed into the lab to work. It wasn't long before he started sleeping later in the mornings and writing later in the day or napping after lunch. Camp Columbia was also growing quieter; nearly half of the garrison had been shipped out to China.

Things were so slow at the camp that Reed began to talk of returning to the U.S. in a few weeks to finish his typhoid report. His presence was not really needed in Cuba at the moment as badly as it was in Washington, where the typhoid report was in its final stages.

Reed was also a little homesick, writing every day about the gardens at his summer home Keewaydin in Pennsylvania, where the strawberries, blackberries and grapes were on the vine. The rhododendrons were in bloom. He missed his favorite foods and mint juleps, instructing Emilie not to let the patch of mint die during the summer heat because he would be very thirsty upon his return. But, Reed had also grown to love Cuba. Red hibiscus bloomed wildly throughout the countryside there, and the umbrella-shaped royal poinciana trees, he wrote, appeared as a "flaming mass of scarlet."

It was the wet season, and rain fell every afternoon, sometimes twice a day, dimming the light in the lab. Steam rose from the railroad tracks in Camp Columbia as the first drops hit the hot metal. And as the storms passed, heading out to sea, lightning would play across the surface of the water, igniting the ceiling of blue-gray cloud cover.

When it wasn't raining, the soldiers at the camp played baseball or went horseback riding. The men worked together all day, only breaking to go to the mess hall. Most men at the camp, twenty years younger than Reed, ate hearty amounts of the meat shipped from the U.S. and local fruits and vegetables. Reed, on the other hand, was more careful with his diet, often joking, "Boys, have mercy on your poor kidneys, they can't be replaced." Reed had adopted the latest trend in Victorian America: healthy eating habits. Large, meat-heavy breakfasts that could include steak, bacon, eggs, sausage, potatoes, porridge and doughnuts were evolving into lighter, healthier meals. Fruits and vegetables replaced primarily potato side dishes. Fresh fish, especially red snapper, were plentiful in Cuba, prepared by the camp's Chinese cook. But Reed was also careful of his diet because he seemed to suffer from stomach ailments—though no one paid much attention to it at the time.

For the most part, the men at Camp Columbia enjoyed Reed's company and he theirs. Practical jokes were common, as were games and sports. The overwhelming opinion of Reed was that he behaved exactly as a major should behave: He was approachable, had a good sense of humor and a genuine concern for the men, but felt a duty to maintain the respect afforded the position. But, as an officer, Reed also stood apart from the men. He could seem uncompromising at times, even rigid.

One contract surgeon described Reed: "He was smart and knew about everything, but he was a man to whom you had to prove what you said . . . he was domineering . . . a good friend, but he insisted upon you being worthy of his friendship."

Reed was not the only one growing impatient with the sleepy pace of work at Camp Columbia. Lazear wrote to his wife that the Quemados epidemic was over, but he would at least now have the opportunity to focus on some of his work with mosquitoes. Since he had been assigned to explore the mosquito theory, Lazear had a lot of time to work independently, traveling back and forth to Havana, where yellow fever outbreaks persisted. Lazear was also sent by Doherty wagon into the countryside whenever outbreaks occurred. He continued adding to his collection, taking the insects from inside the netting around hospital cots, caging them and returning to the lab at Camp Columbia.

Lazear also sounded homesick in his letters home—it had been three months since Mabel and Houston sailed for the U.S. He worried about how little Houston would react to his father after not seeing him for months. Mabel was suffering complications with the pregnancy and had already been admitted to the hospital for the final weeks. Lazear began to talk about what sort of work he might pursue in Washington once his service was over. He was having his mother's portrait painted and looked forward to the day he would have a permanent home in which to hang it. Lazear was very close to his mother. Twice widowed, his mother had also lost both of Lazear's brothers. Jesse Lazear was all she had left. He began to make plans to sail home for a visit in October.

Reed sat at his desk one afternoon in July to write to Emilie. A photo of their daughter Blossom was propped up before him. Reed had just returned from Havana, chased back to the barracks by a heavy storm settling in for the afternoon. In Havana, Reed had seen his son, Lawrence, also stationed in Cuba, and he wrote Emilie to tell her about their meeting. Lawrence looked cool and collected, wrote Reed, with a helmet perched on his head. Army life suited Lawrence far more that academics did, and Reed was proud to see his son come into his own. At their home in Washington, Reed had noticed his son's lack of interest in studies and reproached him, refusing to send him to college. Lawrence Reed would not disappoint his father; by the end of his forty-two years in the army, Lawrence would become a major general.

In Havana, Lawrence told his father that he had just bought a khaki suit and a linen crash one for the warm weather. Reed told him to buy a second crash suit, and he would foot the bill. Lawrence smiled. "I will not neglect to do so."

The only real excitement the board had thus far was a visit from two English doctors, Dr. Herbert Durham and Dr. Walter Myers, who were part of the Liverpool School of Tropical Medicine. Reed wrote to Emilie: "I enjoyed meeting them very much, as they were two of the most typical English men, 'don't you know,' that one could possibly meet. We have placed our laboratories at their disposal during the ten days they will be in Havana." The doctors also met Dr. Henry Rose Carter, an American doctor from the South who had made some interesting observations about the incubation period of yellow fever, most notably the five-to-seven-day time span between the first cases of yellow fever and the next wave of infections. And the doctors spent some time with Dr. Carlos Finlay in Havana as well.

The British doctors, like the American Yellow Fever Board, were intrigued by the idea that yellow fever might be spread by an insect. They later published an article in which they thanked the Americans for their hospitality in Cuba. In it, Durham and Myers wrote, "The suggestion propounded by Dr. C. Finlay, of Havana, some twenty years ago, that the disease was spread by means of mosquitoes hardly appears so fanciful in the light of recent discoveries." Six months later, both Durham and Myers contracted yellow fever during their studies in South America. Durham recovered, but Myers died.

As Reed wrote to Emilie that July afternoon, Carroll and Lazear worked in the lab, but Agramonte had been sent into the countryside to investigate a fever outbreak. Again, Reed was thinking of when he could sail for the U.S. and finish his typhoid work. Then, a telegram arrived.

Agramonte, inspecting a camp at Pinar del Rio, the capital of Cuba's westernmost province, had found something. Reports surfaced about a sudden increase of sickness among soldiers. The fevers had an unusually high mortality rate, and on July 19, Agramonte was dispatched to investigate the outbreak. The recent death of a soldier, supposedly suffering from malaria, afforded Agramonte with a fresh autopsy. He began his investigation of the dead, following the internal clues of gilded organs and tar-like blood remnants, searching the bloodstream for malarial parasites. But the yellow corpses left little doubt as to what had caused the death.

Agramonte walked the hot corridors of the military hospital where he found, in bed after bed, patients burning with fever. No quarantine or disinfection measures had been taken. He sent a telegram to headquarters to the chief surgeon describing the gross negligence on the part of the doctors in charge of the hospital and asked for direction. He received a telegram in response with instructions: *Take charge of cases. Reed goes on morning train. Wire for anything wanted. Nurses will be sent. Instructions wired commanding officer. Other doctors should not attend cases. Establish strict quarantine at hospital. You will be relieved as soon as an immune can be sent to replace you. Report daily by wire.*

Reed left Camp Columbia immediately for the western interior of Cuba, traveling by train through a valley of tobacco and sugarcane fields, the smell of molasses thick in the air. Pinar del Rio boasts the most dramatic topography in Cuba. Giant limestone formations, remnants of the Jurassic era, tower over the fields like stone haystacks. A low-lying spine of mountains stands in the distance, and pine trees rise like spires in the countryside.

Reed met Agramonte, and together, they surveyed the sleeping quarters of the barracks where hundreds of beds stood in regimented rows. In such close quarters, the fever should have spread rampantly from one soiled bed to the next. Conditions were filthy, with ample opportunity for germs to spread. The doctors at Pinar del Rio performed only the barest practices of disinfecting bedding, soaking sheets and pillowcases in bichloride, then sponging the mattress. Reed and Agramonte placed all blame for the fiasco on the army doctors stationed there and filed a formal complaint with the surgeon general.

Before he left, Reed noted one other peculiar circumstance at Pinar del Rio. In early June, a private had been court-martialed and sentenced to three months of hard labor. During this time, he was kept in a cell in Pinar del Rio. The private and seven other men were locked in their cells on June 1 and had no contact with the outside world. No visitors saw the prisoners. Nonetheless, by the end of June, the private and one other cellmate had come down with yellow fever and died. None of the other men in the cell, nor any of the guards, became ill. It was as if something blew through the bars of the windows to afflict only those two men.

Reed met with Lazear and Carroll when he returned to Camp Columbia, while Agramonte continued on the train to Havana. Later, much would be made of the fact that Agramonte was not present at that meeting. In an article published several years later, James Carroll would accuse Agramonte of hardly being involved with the board's work and not even being aware of their experiments. But, Agramonte's lab was located in Havana, as it had been long before the Yellow Fever Board was convened. It was not unusual at all for Reed to meet with Carroll and Lazear at Columbia Barracks while Agramonte continued his work elsewhere. And, most likely, Reed had already discussed his new ideas with Agramonte at Pinar del Rio. As Albert Truby remarked, Reed enjoyed talking about his work: "He had no secrets about his experiments and everyone knew from day to day just what was going on."

During that meeting, it was decided that Lazear would focus more heavily on the mosquito theory, building on the work he had already compiled over the previous months. Since the first outbreak of fever in May, Lazear had been making notations about yellow fever—the symptoms of patients, where the fever broke out and where it turned up next, weather observations and careful study of the mosquitoes local to the area. Lazear still believed that if malaria could be spread by mosquitoes, it was possible that yellow fever could as well. Carroll and Agramonte would continue their autopsies and tissue samples in search of something in the body—whether microbe or otherwise—that appears in yellow fever patients.

The lab work would be helpful, but most likely inconclusive; Reed knew that the next step would have to be human experiments. In a letter to Sternberg, Reed wrote, "Personally, I feel that only can experimentation on human beings serve to clear the field for further effective work." He never indicates in letters or otherwise whether he believes the mosquito theory will be proven or disproved by such action, only that "with one or two points cleared up, we could then work to so much better advantage." Reed took his idea of human experiments to the board, and they agreed that should they be willing to experiment on human beings, they should be prepared to experiment on themselves as well. How could they ask for the self-sacrifice of others without submitting to the same studies?

Agramonte's departure from the board that August would not be the only one noted. According to James Carroll, Walter Reed left without explanation the morning after the board decided to self-experiment and sailed for the United States. Carroll overtly implied that Reed did so to avoid infecting himself. It became a stain on the board's work and indirectly accused Reed of cowardice. In truth, whether or not Reed planned to infect himself, he had been plotting his return to the United States weeks before this meeting took place. In letters to Emilie, as well as Sternberg, Reed wrote about plans to catch the *Rawlins* whenever it came into port. What's more, travel to and from Cuba was unreliable at best. Transports could be days late, and baggage was subject to a lengthy disinfecting process before the steamer could set sail. In letters to Emilie, Reed tried to sound reassuring: "The Quartermaster told me this afternoon that the *Rawlins* would be here Tuesday afternoon, & would leave either Wednesday evening or Thursday morning—So that the chances of getting away August 1 or 2nd are very good." In other words, it would be nearly impossible to jump aboard a ship the morning after the board met and flee.

Carroll went so far as to suggest that Reed fabricated a reason to leave Cuba as well, having a carte blanche to sail to and from Cuba whenever he liked. When Reed and Carroll arrived in Cuba that June to work with the Yellow Fever Board, they learned that Dr. Shakespeare, Reed's partner in the typhoid study, had died suddenly from a heart attack. That left only Dr. Victor Vaughan to finish all of the work. Reed wrote on several occasions about plans to return to help Vaughan finish the report, and under some pressure from the surgeon general to present the typhoid paper, Reed undoubtedly felt a priority to that study. After all, yellow fever had been well contained at Quemados, and with the sickly season coming to a close, even autopsies were few and far between. The fact that Carroll would make such a slanderous suggestion about his friend and colleague seems to further illuminate his emotional deterioration during the time he served on the board. For Carroll, the work would soon prove to be physically harrowing as well, bringing him to the brink of death and back again.

The steamer *Rawlins* was scheduled to reach the Havana harbor around July 30. Slightly irritated, Reed waited as all of his baggage was disinfected in the days before the transport arrived. All wearing apparel was steamed or soaked in a disinfectant called formalin. No exceptions were made—even for the head of the Yellow Fever Commission. Then, much to Reed's frustration, the steamer's departure was postponed after several soldiers decided to celebrate their departure at a nearby café. At long last, the *Rawlins,* which Reed nicknamed the *Rollins* in anticipation of his usual seasickness, set sail close to midnight on Thursday, August 2, 1900. Reed wrote to Emilie that he wanted fried chicken, cabbage and waffles upon his return.

Albert Truby boarded the *Rawlins* late that night with the rowdy group of homeward-bound soldiers from the café. He made his way through the drunken shouts and cheers to his stateroom, where he was surprised to find Walter Reed on the lower bunk. In all the confusion, the only two medical officers on board had been given one room until separate quarters could be assigned.

As the lights in their cabin went out that night, Reed asked, "Doctor, were you mixed up in the celebration?" Truby dutifully explained that he had been at the quarantine station all day, but secretly, he was flattered by Reed's concern. "He had always shown some interest in me since my entrance examination," Truby would later write.

During the few days at sea, Reed and Truby discussed yellow fever. Reed was excited by the work the board had done examining the blood of yellow fever victims. Their findings, along with Agramonte's independent blood cultures, had finally put an end to Sanarelli's claims; Reed was satisfied that they'd finished that first, important objective, and he was anxious to see what Lazear had found in his insect work. Now, they could finally turn their attention to the mosquito theory.

As the *Rawlins* made its way north, and the sea air cooled, Reed talked to Truby openly and excitedly about the work on yellow fever. Duty done, he looked forward to returning to Cuba as soon as possible to launch the investigative portion of their study—his real passion. "He was much pleased," wrote Truby, "with the deep interest Lazear was showing in the mosquito work."

Yellow fever had been arguably the most feared disease in America and the Caribbean for two centuries; a few more weeks could hardly make a difference.

CHAPTER 15

Vivisection

Giuseppe Sanarelli's paper on yellow fever had done more than spark a public brawl with Surgeon General Sternberg over the cause of the fever; it spotlighted the protest against vivisection and human experimentation in the late nineteenth century. Sanarelli, bruised by skepticism from Sternberg and the Hopkins doctors, boldly countered that he had managed to produce yellow fever in five hospital patients. His attempt at "scientific murder" brought a firestorm of opposition from the antivivisectionists, as well as the John Hopkins doctors. Sir William Osler spoke out publicly: "To deliberately inject a poison of known high degree of virulency into a human being, unless you obtain that man's sanction, is not ridiculous, it is criminal."

Vivisection, literally to cut open or dissect a living organism, was an umbrella term that not only applied to surgery and autopsy, but all medical experiments—those on animals, as well as humans.

In the late Victorian age, amid macabre tales of Jack the Ripper dissecting his victims and medical students robbing graves to find cadavers for autopsy study, the antivivisection activists rallied behind their cause—to place limitations on the physicians and scientists assuming a Godlike power over the human body, whether it is alive or dead.

As with most causes, the extremes dominated the argument. Antivivisectionists often quoted Alfred Lord Tennyson whose poem "In the Children's Hospital" described the surgeon, with his ghastly tools, who "was happier using the knife than in trying to save the limb." Scientists countered that no advances in medicine could be made without surgery, autopsy and experimentation. The antivivisection movement had even slowed the Typhoid Commission's work, as Vaughan wrote, "To order autopsies would increase the public furor which at that time was running high among the people." What's more, if the antivivisectionists pushed to outlaw all animal testing, they forced doctors to carry out experiments on humans. The activists, vehemently against the mistreatment of animals, had brought humans into the debate as more of an afterthought; in fact, the Humane Society would one day splinter from the antivivisectionist movement.

Human vivisection had a long history, and it would not come to an end until the Tuskegee Syphilis Study in 1972, but it was in its heyday in the late nineteenth century. As one writer observed, "The use of human beings to confirm that a microbe caused a particular disease or to demonstrate the mode of transmission was a harsh legacy of the germ theory of disease." Louis Pasteur and Robert Koch had provided the techniques necessary for studying germs, which included isolating a bug, growing a culture and finally using the germ to generate disease in a healthy organism, most often a human. Physicians routinely infected themselves, their children, unknowing patients, as well as infants, criminals, the dying and the mentally impaired.

Children were often subjects for medical testing since they were essentially clean slates with little exposure to disease. Nearly a decade before Edward Jenner's famous vaccination against smallpox, he infected his ten-month-old son with swinepox. When the infant became ill, he tested him with smallpox at least six times. Jenner's son would remain a sickly and mentally impaired child who finally died at age twenty-one. For the rest of his life, Jenner could not discuss his son without crying. Nearly a century later, Surgeon General George Sternberg and Walter Reed, in 1895, also used children in several orphanages to test a smallpox vaccine. The fact that Sternberg and Reed certainly believed they controlled the experiments and were not putting the children at risk only furthered the idea that they, like God, could conjure disease and cures at will.

One of the notable examples of self-experimentation was that of William Halsted, who tested the anesthetic capabilities of cocaine on himself. He would continue his brilliant career at Johns Hopkins, but suffered a lifelong battle with addiction. George Sternberg had gone so far as to self-experiment with gonorrhea in the 1880s, though fortunately for his wife, he failed to produce a positive result. And even yellow fever had been the subject of self-experimentation in the past. Dr. Stubbins Ffirth, in 1802, attempted to prove the fever was not contagious by injecting himself with tainted blood and swallowing pills made of black vomit. By not understanding the incubation period of the virus, Ffirth unknowingly spared himself a case of yellow fever. Carlos Finlay had been trying since 1880 to prove his mosquito theory of yellow fever, infecting around ninety human subjects, including a group of Jesuit priests, with the blood of fever patients. All had proved unsuccessful.

History's most famous case of vivisection was yet to come however. As the Yellow Fever Board agreed to self-experiment in the following months, they seemed to have little confidence in the studies. At the very least, they believed it would be a long process with at least a year or two of work ahead of them.

Burial of the Dead, a photograph taken during the Spanish-American War. While 365 American soldiers died in battle, 2,500 died of disease, prompting the U.S. government to send medical teams to investigate outbreaks. Major Walter Reed headed the Yellow Fever Commission.

Philip S. Hench Walter Reed Collection at the University of Virginia

Aristides Agramonte, James Carroll, and Jesse Lazear, members of Walter Reed's Yellow Fever Commission at Camp Columbia outside of Havana, Cuba, in August 1900. Both James Carroll and Jesse Lazear allowed "loaded" mosquitoes to bite them as part of their experiments, and both came down with severe cases of the fever.

Philip S. Hench Walter Reed Collection at the University of Virginia

Walter Reed with his daughter, Blossom, at their summer home, Keewaydin, near Blue Ridge Summit, Pennsylvania. Keewaydin, named after the Indian word for "west wind," was a place of sanctuary for the Reed family, and Reed often wrote about it in letters to his wife. This photo was taken in 1900 when Reed was head of the Yellow Fever Board.

Philip S. Hench Walter Reed Collection
at the University of Virginia

Walter Reed in Washington, D.C., 1882.

Philip S. Hench Walter Reed Collection
at the University of Virginia

A snapshot of Jesse Lazear, taken from his personal photo album of Cuba. Other photos in the album include sites around Havana and photos of his wife, Mabel, and son, Houston.

Philip S. Hench Walter Reed Collection at the University of Virginia

Jesse Lazear holding his son, Houston, outside their home at Camp Columbia. Gertrude, Houston's nanny, stands on the porch.

Philip S. Hench Walter Reed Collection at the University of Virginia

Aristides Agramonte, 1902.

Philip S. Hench Walter Reed Collection
at the University of Virginia

James Carroll, 1900.

Philip S. Hench Walter Reed Collection
at the University of Virginia

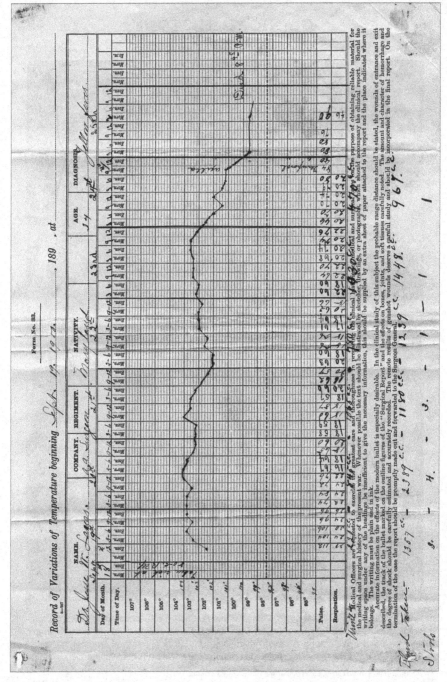

Jesse Lazear's fever chart, September 1900. Lazear was the one member of the Yellow Fever Board who believed wholeheartedly in the mosquito theory. He suffered a terrible case of yellow fever and died on September 25, 1900, at the age of thirty-four, a martyr to science.

Philip S. Hench Walter Reed Collection at the University of Virginia

Detachment of the Hospital Corps, Camp Columbia, Cuba, 1900. In the hand-labeled photograph, No.1 is Dr. Albert Truby, No. 4 is Dr. Robert P. Cooke, No. 10 is Private John Kissinger, and No. 25 is Private John Andrus. Private John Moran is not shown. Fourteen of Walter Reed's volunteers for the yellow fever experiments came from this detachment.

Camp Lazear on Finca San Jose in the Marianao suburb of Havana in 1900. Building No. 2, the Infected Mosquito Building, is in the foreground with Building No. 1, the Infected Clothing Building, in the distance.

Philip S. Hench Walter Reed Collection at the University of Virginia

Building No. 1, the Infected Clothing Building, in December 2005. The building is all that is left of Camp Lazear. Albert Truby and historian Philip S. Hench located Building No. 1 during the 1940s and 50s, working with the Cuban government to restore it to its original condition. After Castro took over, the project was abandoned. Today, the building, which stands in a slum section of Havana, is in a state of disrepair.

Mark R. Crosby

Philip S. Hench had this wall erected beside Building No. 1 to commemorate Jesse Lazear, Walter Reed, James Carroll, Aristides Agramonte, Carlos Finlay, and other key participants in the yellow fever studies. A plaque on the wall lists the volunteers involved in the experiments.

Mark R. Crosby

CHAPTER 16

Did the *Mosquito* Do It?

When the epidemic in Quemados ended, and Reed had sailed for the U.S., the doctors of the Yellow Fever Board found themselves without subjects to research. They decided to move their studies to Las Animas Hospital, literally "the Souls Hospital," in Havana, where there existed a large number of yellow fever patients, but also Agramonte's fully equipped laboratory. The doctors settled into a quiet pattern: Agramonte performed autopsies on the yellow fever cadavers in Havana, Carroll continued his work with tissue samples in his lab at Camp Columbia, and Lazear nurtured his mosquitoes.

To Dr. Carlos Finlay's delight, the Yellow Fever Board had paid him a visit to request samples of mosquito eggs. Finlay was now sixty-five years old, and he had been arguing his case for mosquitoes for twenty years. He took them into his library and pulled a few eggs from the glass edge of a bowl half full of water. Lazear listened intently as Finlay described the life cycle of this particular type of striped house mosquito, its peculiar aptitude for dwellings and fresh-water cisterns, its tendency to stay in a localized area rather than traveling great lengths, the female's persistence in biting at all times of the day, its genius for adapting from its native forests to human habitats. *Aedes aegypti* was also a low flyer, drawn to the human scents of carbon dioxide and lactic acid that were heavier than the air and sank downward toward the exposed flesh of ankles and legs. Then, Finlay gave several dried eggs of that particular mosquito species to Lazear to raise for his own studies.

In addition to the stages of the mosquito's development, there were anatomical details to track. Given Lazear's methodical nature, the care of the "birds" was perfectly suited for him. As the eggs hatched, Lazear carefully labeled the glass tubes and shipped a few more of the mosquitoes to Dr. Leland Howard, an entomologist with the Department of Agriculture in the U.S. Sure enough, the samples proved to be the striped house mosquito so common in North America. As that generation of insects died, Lazear nurtured the eggs of the next. The most essential factor in rearing a batch of mosquitoes and producing new eggs is a fresh blood supply. The pregnant females need blood in order to lay eggs; and once hatched, the new generation relies on blood meals to thrive.

Las Animas Hospital housed an ample source of yellow fever patients. Lazear carried his fledgling mosquitoes in glass tubes plugged with cotton to the yellow fever ward. He removed the cotton and turned the tube upside down against a man's arm or abdomen until the mosquito zigzagged its way downward, legs arched, and struck.

Though each of the members of the Yellow Fever Board had volunteered to self-infect, it turned out to be more complicated than that: Reed's work in the U.S. had detained him longer than

planned as he prepared his typhoid report, Cuban-born Agramonte was thought to be immune, and Carroll spent most of his
time working at Camp Columbia. Only Lazear, who frequented
Havana, was left to test the mosquito theory.

On August 11, Jesse Lazear took one of his carefully labeled
yellow fever mosquitoes, flipped the tube upside down and waited
as the insect landed and bit his forearm. When the mosquito
seemed to have had her fill, Lazear tapped the glass, and she flew
upward again. He marked it in his logbook. Contract surgeon Alva
Sherman Pinto also volunteered to be bitten. Neither resulted in a
case of yellow fever, though the mosquitoes had fed on infected
patients and should have been carrying the virus in their wiry,
winged bodies. Lazear's notebook lists a number of other volunteers from Las Animas Hospital, none of whom developed yellow
fever. The board became discouraged, especially Lazear, who was
ready to "throw up the sponge."

Then, the prey turned on the predator.

On August 23, Lazear sat down and wrote to his wife, who was
now in the hospital on bed rest as she awaited the birth of their
baby. In his letter, Lazear expressed his frustration with the fact
that both Reed and Carroll seemed so preoccupied with the
Sanarelli controversy: "Reed and Carroll have been at that for a
long time and they have notions as to what we should do that I
don't agree with. They are not inclined to attempt as much as I
would like to see done . . . I would rather try to find the germ
without bothering about Sanarelli."

In letters to both his mother and his wife, Lazear never mentions the experiments he conducted on himself. It may be that he
did not want to worry them; two days later, he received word that

his wife, Mabel, gave birth to a daughter named Margaret. He wrote to his mother, "The distance seems very great at a time like this."

On August 27, Lazear finished his work at the lab at Las Animas Hospital in Havana and collected his glass tubes of mosquitoes to take with him to Camp Columbia. It was getting close to noon, the rooms of Las Animas grew stale, and the sun beat down against the tile floor. If he didn't hurry, the afternoon rain would set in during their ride back to camp. Lazear didn't want to take the time to return his glass-caged insects to Agramonte's lab at the military hospital, so he carefully packed up the test tubes in a carrying case to take with him. Cuban doctors watched with curious amusement as Lazear came and went, his arms full of caged mosquitoes, just as Finlay had done twenty years before. As he left the hospital, Lazear noticed that the mosquitoes seemed to be thriving—all but one that is. Lazear made a note that morning that one of the females, who had bitten a new yellow fever patient twelve days before, had refused to feed.

   Lunch ended at the mess hall at Camp Columbia, though the afternoon breeze still carried the smell of food and coffee. Most of the men were heading to their bunks to nap through the rainy part of the day, but Lazear returned to the lab to check on the dying mosquito. Carroll was seated at his microscope. Light poured through windows and cracks of the wooden walls leaving a pattern of lines and squares throughout the lab like a giant white web. Lazear tapped the glass tube and watched the listless female mosquito. He complained to Carroll that she had refused to feed that morning and would likely die by the next day. Carroll rolled up his sleeve and volunteered his arm. Without much thought, Lazear held the glass tube against Carroll's pale skin and waited for the lethargic insect to light on his arm. But the mosquito remained still, clinging to the glass wall. Exasperated, Lazear let go. Carroll took hold of the test tube and patiently held it in place until the mosquito fluttered downward onto his arm and inserted her proboscis like a needle easing into flesh.

At that point, the board had all but given up its hopes of the mosquito as the transmitter of yellow fever. Half a dozen volunteers had fed infected mosquitoes, including Lazear on a number of occasions. Not a single case of yellow fever had developed. Carroll had been skeptical from the start, and by then, believed that the mosquito theory was useless. James Carroll was just feeding one of Lazear's pet mosquitoes to keep it alive; he never expected to get ill.

Two days later, Carroll, Lazear and Agramonte worked in the one-story Las Animas Hospital in Havana. Lazear's sickly mosquito was robust and healthy again thanks to her blood meal from Carroll. Lazear went about his usual, careful routine of feeding his mosquitoes on the infected patients in the yellow fever ward, plugging the test tubes with cotton and making marks in his leather logbook. Instead of taking the train along the Marianao railroad that afternoon, the three doctors left the hospital by Doherty wagon. When they came to the fork in the road, Agramonte hopped down and headed toward the military hospital on foot; Carroll and Lazear continued the ride along the sun-washed road to Camp Columbia. Carroll seemed quiet and distracted.

The next day, on August 30, Carroll and a few of the officers swam in the bright water off the coast of Cuba; sea bathing had become a favorite pastime. The water, as warm as the air, created a strange, seamless sensation as one stepped from the beach into the sea. As Carroll glided through the water, looking toward the shore where the wide leaves of palms flapped in the breeze, he felt an unusual chill. He eased toward the shallow water and waded ashore. The sun had given him a piercing headache as if every ray of light drove a nail into his skull. One of the contract surgeons took one look at the ashen-colored man staggering out of the sea and said, "yellow fever." "Don't be a damned fool—I have no such thing," Carroll grumbled.

News that Carroll showed symptoms reached Lazear at the camp. He was panic-stricken; his experiment may have produced the first case of yellow fever, but it had infected his colleague. Lazear used the camp telephone to call Agramonte in Havana, worry and fear shaking his voice. He explained that he himself had been bitten just two weeks before without falling ill. Flustered, he added that Carroll had held the tube himself when the mosquito fed. Though he said the words out loud and searched for excuses, his voice grew thin.

When Agramonte arrived at Camp Columbia the next morning, he found Carroll twisted over the microscope searching his own blood for the oblong shape of the malarial parasite. Carroll peered through the lens and tried to sound casual as he told him he caught a cold at the beach; but his bloodshot eyes and his pallid skin, beaded with perspiration, shocked Agramonte.

Carroll remained stubborn, finally having to be ordered to the hospital where his illness spiraled, and he soon became delirious. His temperature rose to 105 degrees, and his heart swelled under the pressure. James Carroll was forty-six years old, and yellow fever proved far more deadly in those over the age of forty.

Lena Warner was called in to nurse him. Warner, who as a child had the fever during the 1878 epidemic, knew exactly what Carroll was feeling. Weak and ill, he tried to tell her that he had been bitten by a mosquito before contracting the fever, but up to that point, it was still an unbelievable theory. His desperate attempts to tell her what had happened were dismissed as fever-induced ravings, and Warner made notes on Carroll's chart that he was delirious. She did, however, agree to Carroll's pleas to go by the board's laboratory on her rounds and drop a small bit of banana into the glass test tubes. Then, he gave very specific directions for replacing the cotton in the top of the tube to prevent the mosquito from escaping. Above all else, the insect was not to get out of its glass cage.

In the lab, Lazear and Agramonte continued to search smears of Carroll's blood for parasites, or any substantial clue to the illness. Rain began to fall and would continue to deluge Cuba in the following days as a tropical storm settled over the mountains of Cuba gathering its dark energy. Lazear flipped through the pages of his notebook to the day that the mosquito had fed on Carroll's arm. Since then, Carroll had visited the yellow fever wards at both Las Animas and Military Hospital No. 1, as well as an autopsy room that was so filthy Carroll refused to work until it was thoroughly disinfected. He had been exposed to yellow fever on several occasions, as he had every day over the last two months. As a case study, he was a decidedly poor one. The only way to prove that it was in fact the mosquito was to try the experiment again.

As Lazear stood in the lab with a test tube in each hand trying to coax a mosquito from one glass house to the other, a soldier walked by the doorway and saluted him. His hands full of test tubes and a stubborn insect, Lazear cheerfully answered, "Good morning," instead of returning the salute. The soldier, curious and encouraged by Lazear's approachable manner, stepped into the room.

"You still fooling with mosquitoes, Doctor?"

"Yes," Lazear said, balancing the tubes, end to end, "will you take a bite?"

"Sure, I ain't scared of 'em." The soldier, like most others, found the work of the Yellow Fever Commission fairly amusing. The mosquito theory, just the thought that these tiny insects as frail and inconsequential as lint with wings could transmit illness, seemed ludicrous.

The soldier had never lived in the tropics before and had not left the base for two months; he was the ideal candidate. Agramonte came into the lab and scribbled the name onto a piece of paper: *William E. Dean.* He would also be known as patient XY. Several days later, Dean became the second known case of experimental yellow fever.

The moon had been brush-marked with clouds all night, and by early morning, as Dean's fever climbed, red light rose like embers off the ocean water. The tropical storm that had been shelling the island with rain all week was making its way out of Cuba and heading toward the Gulf of Mexico with much more energy and intensity than it had previously shown. It would claim 8,000 lives in Galveston, Texas, during that September weekend in 1900.

Warner continued to nurse Carroll, relying on many of the same techniques used over twenty years before in Memphis—the patient was kept very quiet, no food or solids could be given, only small sips of water or lemonade. Cold saline enemas were administered. Though acetylsalicylic acid, or aspirin, had just been created at Germany's Bayer company, the loose powder was not yet in general use for fevers.

In the medical chart, Warner recorded his temperature and pulse every three hours and sent urine samples to the lab twice a day. Carroll's health continued to decline, and his wife received a daily telegram reporting the condition of her husband. His fever hovered around 104 degrees, his skin reddened by the heat; but for visitors, the most disturbing part was watching his body writhe and lurch in the bed.

Reed was finally notified by telegram as to what was happening in Cuba during his absence. He immediately wrote to Kean: "I cannot begin to describe my mental distress and depression over this most unfortunate turn of affairs. We had all determined to experiment on ourselves and I should have taken the same dose had I been there."

After a week of delirium and high fever, Carroll's temperature seemed to subside though his eyes remained saffron yellow. His wife received a telegram that day: *Carroll out of danger.* He had shown none of the telltale black vomiting, and the doctors felt confident that he would eventually recover. When Reed heard the news, he telegrammed Carroll.

September 7, 1900
1:15 P.M.

My Dear Carroll:

   Hip! Hip! Hurrah! God be praised for the news from Cuba today—"Carroll much improved—Prognosis very good!" I shall simply go out and get boiling drunk!

   Really I can never recall such a sense of relief in all my life, as the news of your recovery gives me! Further, too, would you believe it? The Typhoid Report is on its way to the Upper Office. Well, I'm damned if I don't get drunk twice!

   God bless you, my boy.

   Affectionately,
   Reed

   Come home as soon as you can and see your wife and babies.

Reed sealed the letter, but before he sent it, he flipped the envelope over and scrawled in his large, curled handwriting, onto the back: "Did the *Mosquito* DO IT?"

With Carroll on the mend, and Dean recovering from his case of yellow fever, the board decided to stop any further experiments. As Agramonte, whose immunity could not be guaranteed, described it, "We felt that we had been called upon to accomplish such work as did not justify our taking risks which then seemed really unnecessary." Besides, with one colleague down and Reed still in the U.S., the Yellow Fever Board couldn't afford to lose another member now that they had their first real break in solving a disease that had plagued people for centuries.

One member of the board did not heed the warning.

Lazear wrote to his wife, Mabel, from the Columbia Barracks on September 8, "I rather think I am on the track of the real germ, but nothing must be said as yet . . . I have not mentioned it to a soul."

The events that followed and the resulting tragedy would be debated for the next five decades.

CHAPTER 17

Guinea Pig No. 1

On September 13, Jesse Lazear sat in the yellow fever ward of the Las Animas hospital in Havana pressing a glass test tube against the abdomen of a bedridden soldier. The patient's skin was the color of smeared iodine, and his body burned from the fever lit within. Dark, wooden shutters were open, and sunlight streamed across the Spanish tile floor of the hospital room as Lazear held the tube steady and waited for one of his "birds" to take a blood meal from the sick patient. *Aedes aegypti* are particularly sensitive to movement and will flutter away at the slightest twitch. As a vector, it makes the mosquito all the more deadly as during this dance of lighting, then lifting off, she'll bite several times in a short time span trying to get a full meal. With each bite, more of the virus is passed into the host's bloodstream.

Somewhere in the room, Lazear heard the wing beats of a mosquito, as many as 500 beats per second. It was a tinny whine in the stagnant hospital air. He saw the insect flickering like candle-light on the edge of his vision, and then he watched it light onto his arm, its legs crooked, and he felt a pinch. Lazear, who stalked, captured and kept mosquitoes with meticulous obsession, proba-bly first thought to imprison the winged prey in one of his test tubes. But, he was still waiting for the other mosquito, confined beneath glass, to feed on the infected patient. If he moved now, the other one would surely refuse to finish.

Regardless of the many things that consumed his thoughts at that moment, the mosquito had its fill of blood and flitted away before Lazear could capture it for his collection. He had not even gotten a good look at this particular insect. It was probably one of the many malaria mosquitoes that hovered around the hospital for fresh supply.

At least that was the story he told his colleagues.

In his logbook, however, Lazear wrote an unusual entry on September 13. In all cases before that, page after page of records, Lazear had used the soldier's name and simply the date he was bit-ten, with no other attention to the mosquito. A one-line entry with a name and date. On that day, however, in his elegant hand, Lazear did not write the soldier's name, but instead wrote: "Guinea Pig No. 1." He went on to write that this guinea pig had been bitten by a mosquito that developed from an egg laid by a mosquito that fed on a number of yellow fever cases: Suarez, Her-nandez, De Long, Fernandez. It was a precise, detailed history that proved beyond doubt that the mosquito was loaded with the virus when it bit the healthy soldier. The guinea pig's name was never used.

For the next few days, Lazear's life continued much as it had over the last few months in Cuba. He fed and cared for the mosquitoes in the lab. He carefully documented in his logbook Carroll's illness, as well as Dean's, recording blood counts every day. He went sea bathing and ate in the officer's mess hall; he read books by the light of a candle before bed.

Then, he began to lose his appetite. He skipped a few meals in the mess hall. He didn't mention it to anyone, nor did he ask to see one of the yellow fever doctors; instead, he worked hard in the lab trying to ignore the oncoming headache.

On September 18, he complained of feeling "out of sorts" and stayed in his officer's quarters. His head pounded, and Lazear decided to write a letter. Maybe occupying his thoughts with more cheerful things would take his mind off the pain. He wrote to his mother: "Dear little Houston must be very cute. How I wish I could see him . . . Mabel is probably with you now or at any rate will be by the time this reaches you. I wish I could be there too . . . Please don't stop writing often because Mabel has come." He made no mention of feeling ill, nor did he ever mention to his mother or Mabel that James Carroll had fallen ill with yellow fever. That night, Lazear started to feel chilled as the fever came on. He never went to sleep; he worked at his desk through the night, trying to get all the information about his mosquitoes organized. By morning, he showed all the signs of a severe attack of yellow fever. The camp doctors made the diagnosis, and Lazear agreed to go to the yellow fever ward. He asked Albert Truby to look after his belongings during his illness.

Lazear was carried by litter out of the two-room, white pine-board house in which he had lived since he and Mabel first arrived in Cuba. His clear blue eyes were alert as the soldiers held him, and he seemed to fully understand what was happening. The house was fumigated with sulfur dioxide, and Truby removed any valuables, including all books and photographs, as well as a small notebook he found in the blouse of Lazear's army uniform.

Lazear was moved into the yellow fever ward at Camp Columbia, occupying the second to the last dwelling. There, Lena
Warner nursed Jesse Lazear, recording his vitals. The surgeon general required a log of fever cases for army record. The book, maintained by the chief surgeon of the camp, detailed a patient's name,
regiment, nativity, age and medical history. There were physiolog-
ical diagrams for a surgeon to draw exactly where a bullet might
have entered and exited the body, and there was a full-page fever
chart. Warner tracked Lazear's temperature as it rose to 103 degrees, and his pulse hovered around eighty.

Carroll had gained enough strength to walk, though he remained very frail. He and Agramonte made their way through
Camp Columbia, past the hospital buildings, across the tracks,
down the row of one-room buildings that kept the yellow fever
victims isolated as neatly as their jars caged the deadly vectors.
Carroll was profoundly shaken by the sight of his colleague: "I
shall never forget the expression of alarm in his eyes." Lazear's di-
aphragm was beginning to spasm, and he knew better than most
that it meant the black vomit would soon begin. Agramonte
would later write about the relief he felt when he finally saw his
friend's eyes grow blank, and Lazear's mind lost hold of his wife,
son and an infant daughter he had never met. Both doctors
pressed Lazear for more information, but all he would say was
that he had *not* experimented on himself. Instead, he told them a
mosquito in the hospital landed on his arm, and he could not
brush it away.

Reed received a cable with the news that Lazear had contracted yellow fever. He wrote to Carroll immediately: "Your letter of Sept. 20th has just been received, telling me of Lazear's
attack—I am now anxiously awaiting Maj. Kean's next cable . . . I
cannot but believe that Lazear will pull through. I hope & pray
that he does come out alright."

On September 25, Reed sent a letter to Kean: "I have been so ashamed of myself for being here in a safe country, while my associates have been coming down with Yellow Jack." He also wrote: "I somehow feel that Lazear will pull through, as he is such a good, brave fellow."

That same day, Lena Warner braced Lazear's arms with all of her weight, shouting for help. Still, he bolted from the bed, darting around the small, frame-wood room as wildly as a trapped insect beating against glass. Two soldiers ran into the ward, pinning Lazear to the bed, tying restraints onto his wrists and ankles.

The Cuban sun beat down against the one-room shack, and Warner sponged his body with iced whiskey and water. She recorded his temperature, which had held at 104 degrees for days, on the chart beside his bed. His eyes were wild.

It was twilight; she could hear the music from the main hall and the drone of soldier's voices in the distance. Inside the clapboard ward, the dim light and smell of death took Warner out of the camp in Cuba to another time, twenty years before, when she lay on the floor of a plantation house outside of Memphis surrounded by the corpses of her family. Her skin burned at the memory. Flies had swarmed around the bodies, and as always, there was the sound of mosquitoes, the hum of their wings flinching in the hot air.

Lena Warner heard Lazear's restless stirring in the bed, felt the intimacy of watching someone sleep. But the quiet did not last. Lazear's body began to lurch, and black vomit roiled from his mouth, through the bar hanging above his hospital cot. He writhed in the bed, and his skin grew deep yellow. His 104 temperature slowly fell, leveling out at 99 degrees, and Jesse Lazear died at 8:45 p.m. at the age of thirty-four.

His arm and leg restraints were removed, and Albert Truby signed the death certificate for the quartermaster: "I have the honor to inform you that Jesse W. Lazear, Acting Assistant Surgeon, U.S.A., died at this hospital at 8:45 pm, Septr 25, 1900." For the diagnosis, he wrote "yellow fever." For the immediate cause of death, Truby left the space blank; a diagnosis of yellow fever was explanation enough.

Plans were made for a quiet burial so as not to affect the overall morale of the camp; but Truby insisted that Lazear receive full military honors. The entire military personnel of the camp, along with many other friends and officers, walked to the cemetery, while ambulances carried the nurses. Everyone wore white uniforms as they surrounded the gravesite like a colonnade and listened to the post band play taps. Though close to fifty mourners attended the service, only one member of the Yellow Fever Board was there: James Carroll. Reed was still in the U.S., and Agramonte had left a few days before on orders from General Wood to gather supplies in New York.

The same day as the funeral, September 26, Mabel Lazear sat in the home of her mother-in-law on Atlantic Avenue in Beverly, Massachusetts. She felt exhausted after recovering from childbirth and looking after a newborn; she had only been out of the hospital for a couple of weeks. The bell rang that morning, and she was handed a cable message. Her first thought must have been that it was from her husband—she had not heard from him in two weeks, and he had plans to return to the U.S. sometime in October. Then, she read the short cable: *Dr. Lazear died at 8 this evening.* It was signed by Kean, who thought that Mabel received word days ago that her husband was ill with yellow fever. She had not.

* * *

Jesse Lazear's logbook, tall, thin, edged in dark red leather, had its last entry on September 13 with the listing of *Guinea Pig No. 1.* Lazear's distinctive handwriting, with flourishing letters and elegant shape, ended that September, and new entries in different handwriting, most likely that of Walter Reed's, did not begin again until December.

Reed retrieved Lazear's logbook and studied it for clues. After careful examination, Reed concluded that Lazear had most likely infected himself and thus died of a medical suicide. Insurance would not be paid out to his family if this proved to be the case, and in all likelihood, that is why Lazear neither listed his name in the ledger nor admitted to self-infecting. Or, perhaps, Lazear did not want his family to know the truth. Mabel wrote to James Carroll asking for the details of her husband's death, pleading that in spite of his passion for medicine, she could not believe her husband would ever deliberately infect himself. After all, he had a wife and two children.

Shortly after the experiments in Cuba ended, Lazear's logbook disappeared from Walter Reed's office and was not seen for fifty years. The second, smaller notebook that Truby found in the pocket of Lazear's uniform was never seen again. The real cause of Jesse Lazear's death was a secret carried to the grave by every member of the Yellow Fever Board.

CHAPTER 18

Camp Lazear

Walter Reed sailed on the *Crook*, by way of Matanzas, into Havana, on October 4, 1900. This time he approached the city with a poignant mixture of sorrow and purpose. He could smell coal, Cuban coffee, fruit fallen from the trees and carbolic acid in the streets: September had seen 269 cases of yellow fever, the worst epidemic Havana had experienced since the start of the Spanish-American War. When he had sailed out of Cuba in August, yellow fever had been an assignment, a challenge he felt equal to as a scientist. Few cases had plagued their camp, and the majority of Reed's offensive had been fought beneath the lens of a microscope. Now, the disease was much more than that. It had taken the life of a beloved friend and colleague and impaired the life of another. The war to conquer this disease was nearing its end, but Reed knew that in order to defeat it, he would have to come dan-

gerously close to the enemy, sending his soldiers into the frontlines.

Even worse, the leader of the Yellow Fever Board had not even
been there for the incredible breakthrough. Reed had never expected the results to come so fast, and he was guilt-ridden that he
had not been present when two of his team members contracted
the fever. He did not even get to attend the funeral of his friend
Jesse Lazear. Lazear's death cast a dark shadow over the success of
the board, and personally, it would make it much harder for Reed
to submit more men to the studies needed to prove their theory.

Once again, Reed had not been alone in his cabin on his sail to
Havana. This time, Robert P. Cooke, a young contract doctor recently out of the University of Virginia, shared his cabin. Reed and
Cooke were already acquainted; Cooke had been one of the doctors in charge of the yellow fever ward at Pinar del Rio that summer when Agramonte and Reed found horrible conditions and an
unreported epidemic of fever in the camp. In fact, Cooke had
nearly lost his contract with the army as a result of Reed's formal
reprimand. The acting chief surgeon had written a letter to
Cooke: "Let this awful experience be a lesson to you . . . While
you are not as culpable as your associates, do not flatter yourself
that the authorities will hold you guiltless."

It was decided that Cooke's mistake was due to youth and in-
experience above all else, and he kept his job. Cooke took the criticism and reprimand without argument, and his modest nature
impressed his superiors. Reed took a liking to him as well on
board the *Crook,* and the two doctors—one just beginning his
medical career, the other approaching the highest peak of his—
discussed the potential of the yellow fever work and the tragedy
of Jesse Lazear. Perhaps Cooke felt some guilt about his mistakes
at Pinar del Rio, or maybe he was inspired by Reed and the mar-

tyred Lazear. As the two men left the ship, those conversations onboard settled like silt into the conscience of Robert P. Cooke.

Once back at Camp Columbia, Reed went almost immediately to visit Carroll. A full month had passed since he first contracted the fever, and Reed was shocked to find Carroll still so weak and depressed. It would take another week for Carroll to be strong enough to travel, and then Reed insisted that he return to Washington to spend some time with his family while he recovered. Agramonte was still on leave in the U.S., and that left Walter Reed without his Yellow Fever Board on the brink of one of medicine's greatest discoveries. Usually slow, methodical and deliberate, Reed was kinetic in those first few weeks—he spread out paperwork and books across his oak mess table and worked nonstop. Reed had been given Lazear's lab notebooks immediately upon his return, and he scoured the pages trying to find the right combinations linking the infected mosquito, when it bit the feverish patient and when it bit the healthy one, to the onset of yellow fever. The biggest question in his mind was why on those three attempts, Carroll's, Dean's and Lazear's, the mosquito had been able to pass the virus, when it had failed in all other experiments. Unfortunately, the cases of James Carroll and Jesse Lazear were useless as scientific evidence—both men had been exposed to yellow fever in various other circumstances that could negate the mosquito theory. The only possible pure case was that of Dean, their patient XY. Dean denied ever leaving the camp, but Reed could not be certain, so he had Albert Truby wait on the veranda and record their dialogue as Reed casually engaged Dean in conversation.

"My man, I am studying your case of yellow fever and I want to ask you a few questions. Before questioning you, however, I will give you this ten-dollar gold piece if you will say that you were off this reservation at any time after you left the hospital until you returned sick with yellow fever." Reed fixed his eyes on Dean, who replied, "I'm sorry, sir, but I did not leave the post at any time during that period." The two men sat down, and Reed listened to Dean's straightforward version of what happened. He later told Truby that he was willing "to risk his own reputation" on the veracity of the story.

In the following days, Reed relied heavily on his assistant John Neate and Private John H. Andrus, who had been assigned to help in Lazear's place. Reed contacted Dr. Carlos Finlay, requesting all of his previously published papers about yellow fever, and sent his driver to fetch them. He handwrote the drafts, then asked clerks to type pages for him. Over an eight-day period, Reed wrote a report, 5,000 words long, outlining the mosquito theory. Once his paper had been typed, he had Truby and the other men mail copies to a long list of people with a handwritten note attached: *Compliments of the writers*.

There was urgency to the matter: The American Public Health Association would hold its annual meeting on October 23 in Indianapolis. Reed had already cabled Washington to ask permission to attend, and Sternberg, ever anxious to report the cause of yellow fever, had made space among the 150 delegates for Walter Reed. It was such a last-minute entry, however, that Reed's name would never even appear in the printed program. Reed wrote to Emilie and told her to expect him by October 18, when he would travel to Indianapolis for the presentation of his paper, "The Etiology of Yellow Fever: A Preliminary Note."

Reed had been given twenty minutes to speak, but was then granted an additional twenty for his presentation. He was explicit in his credit to others involved in the discovery, most especially Dr. Carlos Finlay, the "Mosquito Man," who had been ridiculed by both the Spanish and American press. Reed publicly thanked Finlay for supplying the mosquito eggs necessary to the experiments and cited several of his published articles. The *Indianapolis Journal* called Reed's presentation "fascinating," and the *New York Times* published an article in which one health officer praised Reed's theory. "If the Finlay theory is true," said the officer, "the sufferer from abroad can be made harmless at the cost of a few yards of mosquito netting. He may die himself, but he will not kill others and he will not interrupt the business of railways or steamboats."

Reed had hoped to make some final changes to his study before it went to print, but Sternberg, probably anxious for the Sanarelli camp to read it, prematurely sent the article without Reed's approval to the *Philadelphia Medical Journal,* where James Carroll, Aristides Agramonte and Jesse Lazear were all listed as contributing authors. In the article, Reed discounted Sanarelli and his supposed yellow fever bacteria, calling it merely a "secondary invader." Instead, Reed explained, "The mosquito serves as the intermediate host for the parasite of yellow fever."

Naturally, Reed's article received some criticism, especially from those who believed Sanarelli's germ theory. The *Washington Post,* in particular, was harsh in its opinion of the mosquito hypothesis. In one article, they referred to the Yellow Fever Board, "whoever they may be," as putting forth a theory that is "the silliest beyond compare." While the article seems unduly subjective considering the recent connections between malaria and mosquitoes, it also confirmed what Reed had felt all along: That his professional reputation was riding on only one experimentally produced case of yellow fever. Just before leaving Cuba for the Indianapolis presentation, Reed and Kean had met with General Wood to discuss what actions to take when he returned. Kean wrote that Reed stood before the general, "tall, slender, keen and emotional" and convinced Wood with his "earnest and persuasive eloquence of which he was a master" to use $10,000 to fund a camp for further mosquito experiments.

It would be called Camp Lazear.

Wild and uncultivated, a clearing of two acres stood angled steeply between sea and sun. The ground was far enough from highway travel to discourage wayward visitors, and it was well drained and windswept enough to deter unwanted mosquitoes. It was also an area that had never seen yellow fever.

Walter Reed had returned to Cuba on November 5, 1900, and in his absence, he had Agramonte search out a locale for Camp Lazear. Agramonte was the natural choice—he was the only board member present in Cuba at the time, and he had lived there the longest. The land belonged to an ancestral home, 150 years old, called Finca San Jose in Marianao, and it was owned by a friend of Agramonte's. They would lease part of the property for twenty dollars a month and begin building Camp Lazear. The land also had one other important feature: It was only two miles from the yellow fever hospitals of Quemados and Camp Columbia. If their experiments proved successful, they would need those hospitals.

During construction and the experiments, Reed would often wander over and sit on the front porch belonging to the couple who owned the farm. He told them how he loved Cuba and even talked of taking his wife and daughter Blossom to visit, maybe even moving there once he retired.

While the camp was under construction, Reed turned his attention to building a healthy supply of mosquitoes. Finlay's earlier samples might not be enough to sustain the experiments, and as cool weather approached, fewer mosquitoes would be available.

Reed picked up where Lazear had left off in his entomological studies, contacting Leland Howard in the U.S. with insect samples and questions. Boxes of paperwork covered Reed's desk, as did two large leather volumes—600 pages each—of La Roche's history of yellow fever, published in 1853. If Reed's enthusiasm wasn't immediately contagious among the men, it soon would be. As they sat around tables playing cards or visiting on the veranda in the autumn evenings, Reed would interrupt, "Gentlemen! Listen to what La Roche says about the terrible epidemic in Philadelphia in 1793!" The men would gather in the study and listen to Reed read from mosquito pamphlets and studies. Soon, the hunt for new specimens began. Using large-mouthed cyanide bottles, they collected mosquitoes and studied them beneath a strong hand lens.

One night in mid-November, a tropical storm pounded Cuba. The low sky grew gunmetal gray as heavy winds uprooted trees, tossed tents and shook the wooden buildings. Shutters blew open and papers flew, wet and tattered. Reed's collection of lab mosquitoes was blown out to sea. When the storm subsided, only a few dry eggs remained, and the experiments were scheduled to begin any day. Colleagues tried to convince Reed that warm weather would return soon enough, and a new supply of mosquitoes would hatch, but Reed persuaded his men to hunt new mosquitoes with him. They searched drainpipes, upturned cans, broken containers and even privy buckets to skim the surface for mosquito larvae, "wigglers," as they called them. Along the still surface of water, they collected the black, cylinder-shaped eggs, which could be dried or frozen or hatched immediately. Returning to the lab, Reed, Neate and Andrus picked through the findings, separating "wigglers" from eggs and harvesting a whole new batch of the lyre-marked *Aedes aegypti* mosquitoes.

The storm aside, the construction of Camp Lazear was nearly

complete. Everything about the camp had to be uncontaminated. Wagons carried new tents and equipment, all in their original packaging, to Camp Lazear. Wooden floors were built where seven tents would be pitched. Personnel were carefully chosen based on their impeccable military records and an interest in experimental medicine. They also had to be in perfect health—all of the volunteers but one were under the age of thirty. The men were then quarantined.

Reed himself designed the most critical buildings for the camp with meticulous care. One would be dedicated to the mosquitoes; the other would be used to disprove once and for all the theory that yellow fever could be transmitted by objects, infected clothes or close contact. The entire compound would be enclosed in a barbed wire fence with a military guard to deter anyone from entering or leaving.

Building No. 1 became known as the "Infected Clothing Building." Its tongue and groove wooden frame was twenty feet by fourteen feet with glass pane windows and a solid-wood, double-door entry; the windows and door were screened then boarded shut. Every precaution was made to keep the structure free of mosquitoes and sunlight. Three beds stood in the center of the room, surrounded by crates and boxes still sealed shut.

Reed sought out volunteers for the Infected Clothing Building, and the doctor chosen to lead the group was Robert P. Cooke. Six months earlier, Cooke had nearly lost his job thanks to the reprimand by Agramonte and Reed. He had also neglected a potentially explosive epidemic of yellow fever at Pinar del Rio. Now, both Agramonte and Reed watched as Cooke and two other volunteers entered their first experimental building. Though the other two volunteers would receive $100 each, Cooke refused any compensation.

In modern times, it's hard to understand the mentality that

would lead a soldier into knowingly risking his life for the purpose
of medicine. Soldiers are trained to fight and defend; if any illness
befalls them, it's considered a cruel and unjust turn of events. But
prior to World War II and the introduction of penicillin, soldiers
lost their lives to disease far more than bullets. From the time of
the American Revolution through World War I, a soldier knew his
odds of dying from dysentery, cholera, typhoid, smallpox, influenza or yellow fever were greater than those on the battlefield,
so volunteering for human experiments might not seem as much
of a psychological departure as it would today. After all, a soldier's
duty is to defense, and many men felt that the greatest threat to
the American people lay not in enemy warships or troops, but in
disease.

On the evening of November 30, Cooke and the two other men
entered Reed's carefully crafted building and sealed the solid
wood door behind them. A single stove stood in the one-room
house, and it kept the temperature inside somewhere between 90
and 100 degrees at all times. Impenetrable to light or air, the small
room felt like a furnace. The three men began breaking open the
crates and boxes left in the center of the room. As they opened
the first trunk, the odor was so pungent that the men ran outdoors, hands over their mouths, to keep from retching. After a
few minutes, the three men returned and finished unpacking
boxes full of soiled sheets, covered in vomit, sweat and feces from
the yellow fever ward. They dressed in the filthy clothing that had
been worn by dying patients, they covered their cots in sheets
stained with black vomit, and then they spent the next twenty
nights the same way.

For the mosquito trials, Reed felt less certain about seeking
volunteers. It was one thing to ask Cooke and the men to expose

themselves to the filth Reed was certain could not transmit yellow fever; it was something else all together to ask men to volunteer for the same experiments that had killed Jesse Lazear and had almost taken James Carroll. The other doctors let it be known that Reed would need volunteers, and then they waited.

Many of the doctors at Camp Columbia knew of John Moran's situation. Having been honorably discharged from the army that July, Moran worked as a civilian clerk, hoping to save up enough money for medical school. An Irishman who had planned to join the cavalry at the onset of the Spanish-American War, his interest had turned instead to the Hospital Corps. As one contract surgeon told him, "Moran, any man with enough influence can become a captain, but not a doctor."

Moran was well liked, though considered a little green. When he first arrived in Cuba, the corpsmen made sure to teach the new Irishman some helpful Spanish phrases. "They were words," he later wrote, "that I could not speak today in the presence of respectable, Spanish-speaking dames and expect to get away with it."

Moran received around $100 a month for his work as a field clerk under General Fitzhugh Lee and had been allowed to remain at Camp Columbia in spite of his civilian status so that he could save money for school. On his discharge papers, his character was described as "excellent," his services, "honest and faithful."

Like many of the men at Camp Columbia, John Moran was also well acquainted with yellow jack. He arrived at work one morning to find the desk next to him empty, only to learn that the other clerk had just died of yellow fever. He had also heard the famous story of Major Peterson, which had shocked the military in Cuba and made the rounds in the American newspapers. When the handsome, young Peterson died of yellow fever, his wife feared that she too might have the fever. Just hours after his death, she pulled a revolver from her purse and shot herself. And, certainly, Moran had also been an admirer of Jesse Lazear, whose loss was still felt, poignantly, throughout the camp.

One day, Dr. Roger P. Ames, a contract surgeon, approached Moran. He knew of Moran's financial situation and had even suggested Moran consider his own alma mater, Tulane, once he had enough money saved. Ames brought up the subject casually—did Moran know that Major Reed was offering a bonus for men willing to volunteer for his new experiments? The $500 reward could certainly go a long way toward medical school, and he'd be doing Walter Reed a favor. "All right, Doc, I will sleep over it and let you know tomorrow."

"Neither of us," Moran later wrote, "gave very much thought to a possible death lurking in the background."

That night, Moran talked it over with his roommate, Private John R. Kissinger. Moran decided not only to volunteer but to do so without monetary compensation. His mind was made up. Kissinger tried to dissuade Moran from forfeiting the money—especially since he needed it so badly for medical school. They continued to discuss it all through the night until the early hours of morning, when they decided to tell Reed that they would both volunteer. They wanted to tell him as soon as possible, before doubt weakened their resolve, but they thought it best to wait until Reed had dressed for the day and taken breakfast. Then, they made the short walk across camp toward the smell of coffee wafting from the officer's quarters. Reed met the two men on his porch, "Good morning. What can I do for you?"

In the tense silence, Reed looked at the two strangers—one a

private, one dressed in civilian clothes—and waited for the
tongue-tied men to speak. When they finally explained the reason
for the visit, Moran wrote, "The Major's surprise was complete
and so reflected in his countenance. He never expected such
rapid-fire action as confronted him, there and then, in the persons of two human guinea pigs."

Reed rubbed his palms, one over the other, and was about to
answer the men when Kissinger blurted, "That is not all, Sir. We
are volunteering without the bonus or money award which we understand you are offering." Reed looked confused, even concerned.

"That is correct, Major," added Moran. "We are doing it for
medical science."

Reed quietly told the men that he would gladly accept them
for his experiments, and it was later famously recorded that Major
Walter Reed touched his cap and said, "I take my hat off to you,
gentlemen." In another version, it was said that Reed remarked, "I
salute you." Both Moran and Kissinger denied that an officer
would have said as much to enlisted men, but when Reed's son,
Lawrence, was asked about it years later, he said that it sounded
exactly like something his father would have said. Actually, General Lawrence Reed added, "He would have said, 'Gentlemen, I
salute you.'"

Reed would describe the moral courage displayed by
Kissinger as unsurpassed in the annals of the army of the United
States. And in a written recommendation for Moran, Reed would
write, "A man who volunteered, as he did, without hope of any pe-
cuniary reward, but solely in the interests of humanity and medical science, to enter a building purposely infected with yellow
fever . . . should need no word of recommendation from any one."

* * *

Other volunteers for the experiments were acquired by less noble means. Agramonte, the only board member who could speak Spanish, was sent to the Immigration Station across the Bay of Havana for recruits. He hired roughly ten newly arrived immigrants at a time to work as day laborers at Camp Lazear. Straight off the boats, the immigrants were delighted to find easy work picking up stones in a field. They were given bountiful meals and tents to sleep in. Surely, they even observed with appreciation the folds of mosquito netting placed thoughtfully over their beds. They may or may not have noticed the casual, but persistent questioning that came from the officers in the camp: Where did their families originate, had they ever lived in Cuba or the tropics before, had they contracted any diseases—any fevers—since their recent arrival? Did they have a wife or children dependent upon them? If any of the immigrants were underage or had previously lived in the tropics, they were sent away from the camp immediately. At last, when the immigrants had grown comfortable to this life at Camp Lazear, the idea of the experiment was put before them. If the men agreed, they would receive $100 in gold and another $100 if they caught yellow fever, in which case they would be given the best possible care. "Needless to say," wrote Agramonte, "no reference was made to any possible funeral expenses."

The board justified their means by arguing that Spanish immigrants in Cuba routinely expected to suffer a case of yellow fever—they would at least be paid for it this way. As crude as their methods seem to modern science, this was still during the height of vivisection. After all, two of the board members had infected themselves in the course of the last few months. What would make this experiment different, and set a precedent for all future human experiments, was the consent form. Most human experiments in the past had been conducted on unsuspecting patients or under false pretenses. Walter Reed insisted that these volunteers understand fully what they were risking and be compensated for it. Years later, Major Randolph Kean would proudly hang a copy of the bilingual consent form on his office wall.

Once the consent form was signed, Moran and the Spanish volunteers went into isolation on November 20. Kissinger, who had been on the premises for a full month, did not need to be quarantined and was ready to begin the experiments immediately. Just as Lazear had done two months before, Reed set a glass tube against Kissinger's skin and allowed the loaded mosquito to land and bite. Now, they would wait to see if and when Kissinger might develop the fever.

Kissinger earned a sort of hero status around the camp. When he entered the mess hall one day, the soldiers rose from their seats and shouted, "Attention! Hero Kissinger is here!" They continued with the meal, occasionally asking someone to "Pass the hero the butter" or "Pass our hero the pickles." One of the soldiers later wrote that Kissinger "was blushing a rosy red, and was so embarrassed he couldn't eat. For he knew that this was the Army Way of hiding the real sentiment we all felt by making a joke of our deep appreciation we would not express in words."

The days seemed slow going, the experiments uneventful. The men were not allowed to leave the isolated farm, and instead played poker and performed light duties around Camp Lazear, adhering to the old army saying, "If it's there, pick it up; if you can't move it, paint it; if it moves, salute it!"

Moran continued to work as a clerk, this time for Reed. He spent four or five hours a day typing reports—with his two index fingers only—on the army's standard Hammond typewriter. A perfectionist, Moran was even known to retype whole pages if he found an erase mark. Reed once tore up a whole page on yellow fever when he found an error, telling Moran that "perfection is next to impossible on a machine."

By the end of November, Kissinger, Moran and the other volunteers had all been bitten at one time or another, sometimes even twice, but there was not a single case of yellow fever among them. What the virus accomplished in nature with relative ease proved much more complicated in the lab. Like Finlay before them, their mistake lay in the simple matter of timing, and as one historian put it, their harvest would soon ripen with a vengeance.

On December 5, Kissinger volunteered to be bitten for the third time, this time by five different mosquitoes. At least one of the loaded insects in the group had fed on a yellow fever patient in the first three days of fever. The virus slipped silently into his bloodstream, and three days later, on December 8, Kissinger came down with a sudden and strong chill. He shivered as his fever climbed almost immediately to 102.5. His head began to pound, and he felt as though his bones had literally been crushed. He would later describe his bout with yellow fever, "My spine felt twisted and my head swollen and my eyes felt as if they would pop out of my head, even the ends of my fingers felt as though they would snap off." By the time his illness ended, Kissinger's weight dropped to 118 pounds, and he would remain an invalid for the rest of his life, eventually becoming mentally ill. Years later, historian Philip S. Hench wrote that Kissinger "really 'died' physically and mentally when he took sick. Thereafter he has lived only one role—that of a yellow fever martyr and hero."

"The case is a beautiful one," Reed wrote, "and will be seen by the Board of Havana experts today, all of whom, except Finlay, consider the theory a wild one!"

Reed also wrote to Emilie: "Rejoice with me, sweetheart, as, aside from the antitoxin of diphtheria and Koch's discovery of the tubercle bacillus, it will be regarded as the most important piece of work, scientifically, during the 19th century."

In the coming week, Reed produced three more cases of yellow fever among the volunteers.

A haunted stigma began to surround Camp Lazear. A number of the Spanish volunteers fled, refusing any additional experiments. Rumors in Havana circulated that some soldiers had found an old limekiln filled with the bleached bones of Walter Reed's yellow fever volunteers.

To the three men living in the Infected Clothing Building, the news of Kissinger's case only added to their claustrophobic sense of fear, which was already cloaked in isolation and filth. If Reed could produce yellow fever under such sanitary conditions, what chance did the three men enclosed in a tomb of germs have? The men barely slept at night and began to imagine fits of fever and chills.

Just when they were at the psychological breaking point, a fresh box from a fatal case of yellow fever arrived. The box had been sealed shut for days, and when the men finally opened it, they ran out of the building into the dark, where one vomited uncontrollably. At last, on December 19, Cooke and his two volunteers were released from their twenty-night stay in Building No. 1, and the next group of volunteers entered. No cases of yellow fever ever developed from the Infected Clothing Building.

By the end of these experiments, Reed had irrefutable proof that yellow fever could not be transmitted by "germs," infected clothing or air. He had exposed his men to every type of filth for up to twenty days at a time, and not one had contracted the fever. It toppled once and for all the prevailing theory that yellow fever could be spread by filth.

On December 21, John Moran entered Building No. 2, the Infected Mosquito Building. Much like its sister structure a few yards away, Building No. 2 had been carefully constructed by Reed. It was the same size and general make, but instead of one room, this building was split through the center with a finely woven wire screen, and the walls had not only been sealed, but also lined with cheesecloth. Again, like its sister structure, there were three cots, but these were outfitted with pristine, steamed sheets. Cot A stood on the "infected side" of the room, while cots B and C were located on the "safe side," protected by a wall of wire that separated the two. The entire building had been thoroughly disinfected. Two other volunteers stayed on the "safe side" of the wire partition, where they would sleep for over two weeks—they were the experiment's control group.

Reed stood to one side of the screen, his face obscured by the metallic mesh, and watched as John Moran entered the room. Moran was fresh from a bath and wearing nothing but a nightshirt, which he removed before lying down on the cot. He rested on his back, his arms at his sides, and held the mirror Reed had given him. Moran could hear the whine of mosquitoes in the room. He lay very still as the insects began to sense carbon dioxide and flickered closer to the source. He watched his reflection in the hand mirror as the virulent mosquitoes landed on his face and fed. Over a series of similar trials that afternoon, Moran received fifteen swollen bites.

On Christmas Day, four days after he entered the Infected Mosquito Building, John Moran awoke with a headache and a chill. He had made a wager with a fellow soldier that he would be present at lunch, and being an Irishman, he wasn't going to give up so easily. Moran knew that eating heavy amounts of food during a case of yellow fever could have disastrous results, so he picked at his Christmas lunch, trying to hide how little he ate.

"Guess you win," admitted the soldier. "Damn it, you can't kill an Irishman anyway." The soldier reached into his pocket to retrieve the money.

Moran looked up from his plate, "Well, I guess you won and lost."

The soldier's smile faded, creases settling around his eyes and brow. "You don't mean to tell me that you have it, Johnny?"

By 3:00 that afternoon, when Reed arrived, Moran's temperature was 103. Reed stood before him with a broad smile. "Moran, this is one of the happiest days of my life."

Moran's temperature would continue to rise to 104 degrees, as he ate nothing but cracked ice and sipped strained watermelon juice for two weeks. During the course of his illness, he lost twenty pounds before making a full recovery.

Though they occupied the same building, ate the same food and breathed the same air as John Moran, the volunteers from the "safe side" of the building never came in contact with mosquitoes. Neither man ever contracted yellow fever.

In the week before Christmas, the atmosphere in Havana was festive, almost like a carnival. In spite of a toothache and no dentist available to treat it, Reed took part in the festivities. He walked the gaslit streets of Havana on the night of December 22, where porch fronts were studded with tropical flowers instead of evergreen boughs. Poinsettia bushes were in bloom, and candles lined windows and church doorways streaming wax down the stucco.

Son music was played in taverns and cafes, the sound of the guitar and flute rising in the air with the beat of maracas like shifting sand beneath it.

Reed crossed Parque Central in the heart of Havana, walking beneath the almond trees and iron lampposts. The triangular roof of the Tacón theater rose above the treetops. He waited for the horse-and-buggy traffic to slow before crossing the street at the Inglaterra Hotel, its neoclassical façade a perfect series of windows and wrought-iron balconies. Next door, between the Inglaterra and Hotel Telegrafo, he walked beneath white columns and archways to the doorway of a narrow building. Even from the street, he had heard the sound of cocktail-laced laughter and smelled the cigar smoke from the open, floor-to-ceiling windows. His dress shoes clapped against the tile as he climbed the narrow case of thirty-one stairs and entered the dining salon of Old Delmonico's Restaurant. The room trembled with candlelight reflecting off crystal, the voices drowning out the music and street traffic below. The dinner was in honor of Carlos Finlay and his mosquito hypothesis; all of the speeches were given in Spanish. Juan Guitéras, the longtime yellow fever doctor who had served both on the 1879 Yellow Fever Commission and as the doctor treating Victor Vaughan at Siboney, was the master of ceremonies. He compared Finlay to Sir Patrick Manson, who first proposed that malaria could be spread by mosquitoes, and Reed to Ronald Ross, who had given the final proof. Carlos Finlay was given a bronze statuette in honor of his valuable work. He would also be nominated for the Nobel Prize in 1905, 1906, 1907, 1912, 1913, 1914 and 1915 for his work with yellow fever, though he would never win. And to this day, Carlos Finlay is Cuba's most revered physician.

At close to midnight, Finlay raised his glass: "Twenty years ago, guided by indications which I deemed certain, I sallied forth into an arid and unknown field; I discovered therein a stone, rough in appearance; I picked it up . . . polished and examined it carefully, arriving at the conclusion that we had discovered a rough diamond. But nobody would believe us, till years later there arrived a commission, composed of intelligent men, experts in the required kind of work, who in a short time extracted from the rough shell the stone to whose brilliance none can now be blind."

There was applause, and glasses were raised. Brandy was poured into cups of coffee, and match tips lit the Partagás cigars. All of the important names were in attendance, including Kean, Agramonte, and of course, Reed. The only person not in attendance was James Carroll, who had returned to Cuba in mid-November. Carroll wrote to his wife that he was "ashamed to go and be the only person present in Khaki which is intended only for a field uniform." As a contract doctor, he was entitled to an officer's uniform, but had never been able to afford one. The tone of this letter, and others like it, betray more than wounded pride. Since his bout with yellow fever, Carroll had become bitter. As a noncommissioned officer, he was slighted by senior officers, overlooked for promotions and underpaid—all of this in spite of the fact that he had nearly died in his service in the Army Medical Corps. Carroll also felt that he had been overshadowed by the success of Reed. In the coming years, James Carroll would only grow more bitter.

For Christmas itself, the wives of Majors Kean and Stark threw a celebration in town. They trimmed a guava bush and handed out gifts. Walter Reed was given an oddly shaped present. When he opened it, laughter erupted. It was a makeshift wire mosquito with a note attached:

Over the plains of Cuba,
Roams the mosquito wild,
No one can catch or tame her,
For she is Nature's child.
With Yellow Jack she fills herself,
And none her pleasure mar,
Till Major Reed does capture her,
And puts her in a jar.
And now alas! For Culex,
She has our sympathy-y,
For since the Major spotted her,
She longs to be a flea.

CHAPTER 19

A New Century

Columbia Barracks,
Quemados, Cuba,
Decr. 31st 1900.

My precious wifie:

*11:50 P.M. Dec. 31st 1900 — Only 10 minutes of the old Century re-
main, lovie, dear. Here I have been sitting reading that most wonderful
book — La Roche on Yellow fever — written in 1853 — Forty-seven years
later it has been permitted to me & my assistants to lift the impenetrable
veil that has surrounded the causation of this [most] dreadful pest of
humanity and to put it on a rational & scientific basis — I thank God
that this has been accomplished during the latter days of the old
century — May its cure be wrought out in the early days of the new century! The prayer that has been mine for twenty or more years, that I
might be permitted in some way or sometime to do something to allevi-*

ate human suffering, has been answered! 12 midnight! A thousand happy new years to my precious, thrice precious wifie and daughter! Congratulations to my sweet girls on their good health upon the arrival of the New Century! Hark! there go the 24 buglers, all in concert, sounding "Taps" for the old year! How beautiful it floats through the midnight air and how appropriate! Good-night my sweet joys, a thousand sweet dreams of father and dear brother! kisses & love & love & kisses for my precious, thrice precious girls in these first minutes of the 20th Century!

Devotedly,
Papa

CHAPTER 20

Blood

Just before the New Year of the new century, the board began injecting blood taken from yellow fever victims into volunteers. With trepidation in his tone, Reed had written to Surgeon General Sternberg to ask if he would still like the board to conduct blood experiments. At that point, Reed was certain yellow fever was in the blood and transmitted through the bite of a mosquito. Injecting patients with yellow fever–thick blood would undoubtedly produce more cases, and thus far, the board had been lucky enough and vigilant enough to nurse all of their yellow fever cases back to health. Not one fatality had occurred. This was probably due in part to the fact that only healthy, young volunteers were accepted, but it was also thanks to Dr. Roger Post Ames, the camp's yellow fever specialist who had, by far, the highest success rate in nursing victims of yellow jack back to health. His success may have been in his psychological approach to patients. He and his male nurse, Lambert, were known to talk away the fear and panic that often came with the fever. He instilled confidence in the men and once told a group of soldiers that, "No doctor should ask a man to submit to a disease unless he knows he can cure him." Ames was from New Orleans and had survived the 1878 epidemic as a child; he believed that the mild childhood case had given him immunity. He was wrong.

With a resounding "yes" from the surgeon general to continue to the final phase of experiments using blood, the board planned to use four volunteers who would receive gradually decreasing doses of infected blood. The blood also had to be "ripe," having been drawn from a patient in the first few days of fever. The word *ripe* seems ironic given that it usually implies harvesting something that has grown sweet, full and fleshy. Blood, the scientists would find, was indeed ripe; but the fruit of their harvest would instead be disease.

The first volunteer received the full dose—2cc—of ripe blood from one of Reed's yellow fever subjects. Within four days he was stricken with a nonfatal case of yellow fever. Then, 1.5cc of blood was drawn from his veins and injected into the second volunteer. In two and a half days, the second volunteer came down with a "pretty infection," according to Reed. The blood flowed from one volunteer to the next, the yellow fever virus floating in it like leaves swept in a current.

After that, the board's experiments stalled while they waited for additional volunteers. At last, two more came forward. A private from the Hospital Corps was injected with 0.5cc of ripe blood taken from a recently fatal case of yellow jack and fell feverish two days later. Four control subjects, Kissinger, Moran and two Spaniards, all of whom had suffered previous cases of yellow fever during the experiments, were injected with ripe blood. None of the control subjects contracted the fever, proving that their previous cases had provided immunity. All that was needed now was the fourth and final blood experiment, and Reed could return to his lab in the States to dissect mosquitoes and begin hunting for the agent causing yellow fever.

On the morning of January 24, 1901, John H. Andrus reported to work in the board's lab as usual. He was a twenty-two-year-old member of the Hospital Corps who had been assigned in October to work with Reed, Carroll and their assistant, Neate, in the lab. Andrus's job, like Jesse Lazear's had once been, was the raising and caring for the lab's mosquitoes—their "pets." Andrus captured the mosquitoes from stagnant water in epidemic-ridden areas and raised them in the jars of the lab, even allowing the females to bite him when blood was needed for the next generation of eggs.

Andrus was busy with his pet mosquitoes that morning when Reed and Carroll entered the lab in mid-argument. Neither seemed to acknowledge Andrus as they bickered about the blood of one of their volunteers. The blood was ripe, and the window of time for injecting it into a new volunteer was narrowing, but the board had just learned that their fourth volunteer had backed out. Reed was frustrated and impatient. He had already waited two weeks for the final volunteers to come forward, and he was anxious to finish the study. Reed decided he would be the fourth volunteer.

Carroll pleaded with him not to—as a test subject, Reed could not have been a poorer choice. He was thirty years older than most of the volunteers and was continuing to suffer from some kind of stomach distress. Inoculating himself would essentially be suicide. Carroll himself, still struggling to recover from his bout of yellow fever months before, had barely survived. Lazear had already been lost.

Reed was convinced that by standing in as the fourth volunteer, he could also prove or disprove a new theory of Carlos Finlay's. Finlay believed one generation of mosquitoes could transmit yellow fever to the next. Reed had been bitten repeatedly by second-generation mosquitoes without infection, so he was either immune to the fever, or Finlay's experiments were wrong. What better way to prove both theories than this final blood experiment?

Andrus listened to the two men argue throughout the day until, at last, Reed's temper flared. As Reed left the lab, he declared that he would be inoculated tomorrow, and there was no use in discussing it further.

The next morning, Andrus returned to the lab to find Carroll, as usual, at the microscope on the long, wooden lab table. The morning sun brightened the lab, lighting the room and giving a silver patina to the collection of vials, jars and test tubes. John Andrus nervously approached Carroll and asked if he himself might take the inoculation in place of Major Reed. Carroll looked up from his microscope, impressed and disappointed, and answered no. Reed was determined to test Finlay's theory as well as his own, and no one else would qualify for both experiments. Andrus reminded him that he had been feeding second-generation mosquitoes for weeks in the lab to keep the females alive; he, like Reed, could perform both experiments at once.

When Reed arrived at the lab that morning, Carroll sent Andrus on an errand, so that the two men could talk. Andrus never knew what was said, but when he returned, Reed asked if he realized what he was doing. Andrus had, in fact, nursed a number of cases of yellow jack, and he had seen many men die of it—but he had no noble intentions of saving Walter Reed's life by risking his own. He later wrote: "I knew something of what proof of the *mosquito theory* would mean to humanity. I knew that Major Reed was the main spring that made the work of the Board tick and that if he was sick of yellow fever and had a slow recovery . . . the work would all but stop."

By 12:15 that afternoon, Andrus was seated in a chair with his sleeve rolled up. The doctor swabbed the skin on his arm with an alcohol-soaked cotton ball and slid the needle in, slowly plunging 1cc of ripe blood. "I knew that, from the instant that needle pierced my skin," Andrus wrote, "no power on earth could prevent my getting yellow fever."

Soon after the experiment, John Andrus was sent by ambulance to Camp Lazear to await further results. He was assigned to a small tent and allowed to roam the camp, while he waited for the virus to take up residence in his bloodstream. One of the first things he did was write to his mother that he had been detailed to accompany a cavalry troop into the interior of Cuba, where he would not be able to send mail again for the next three weeks. Adding to his anxiety was the fact that Dr. Roger Post Ames, the doctor solely credited with nursing every volunteer back to health during the yellow fever experiments, was himself sick with yellow fever.

Three days later, alone in his tent, John Andrus felt a chill come on. He had complained of a headache for the last couple of days, but had shown no signs of fever. His temperature soared to 103.6 degrees. Andrus's face was flushed, and his eyes grew glassy. He remembered being lifted into the two-mule ambulance that would carry him to the yellow fever hospital, but he never remembered arriving there.

The official report states that by the time John Andrus arrived at the hospital, his temperature was over 104 degrees and his pulse raced at 120. Restless, he complained of bright lights, a searing headache and a backache. He vomited several times. Andrus had a severe case of yellow fever.

Reed wrote to the surgeon general, "Should he die, I shall regret that I ever undertook this work. The responsibility for the life of a human being weighs upon me very heavily just at present, and I am dreadfully melancholic. Everything is being done for him that we know to do." Reed's guilt must have consumed him. He had voluntarily allowed another man to stand in his place, to exchange his life for Reed's.

Eleven days after the initial attack, John Andrus was making a slow recovery, but with complications. Dr. Roger Post Ames heard of the case from his own sick bed. After all nurses had left his station, Dr. Ames ordered his assistant, Lambert, and another man to carry his own cot to the yellow fever ward where he could examine and hopefully help Andrus. With time, and the help of Dr. Ames, Andrus recovered fully.

Forty years later, John H. Andrus would lie in Walter Reed Hospital in Washington, D.C., where he would spend the rest of his life bedridden due to spinal problems he attributed to his case of yellow fever. Though doctors could never find the true cause for his paralysis, Andrus wrote, "I would rather think my condition is due to my humble part in the work of Major Reed and his Yellow Fever Board than from some mysterious and unaccountable cause."

The final blood experiments answered a number of questions about yellow fever. Not only was the virus living in the blood of mosquitoes but also in that of humans. Their discovery explained why an epidemic could plague a city for an entire season. And, though the people may die off or flee, the loaded mosquitoes only grow more prolific until the air itself seems steeped in poison.

Nature had found the perfect place to hide the yellow fever virus. It seeded itself and grew in the blood, blooming yellow and running red.

CHAPTER 21

The Etiology of Yellow Fever

With a tribute to Jesse Lazear, Reed presented his paper "The Etiology of Yellow Fever—An Additional Note" to the Pan-American Medical Congress in Havana in February of 1901. In it, he outlined his experiments with the infected clothing house, the mosquito house and the blood inoculations. The hall was packed as Reed read his paper, and listeners even crowded the doorways to hear the monumental news about a disease that had plagued both Cuba and America for two hundred years.

Reed had a habit of emphasizing important points by raising his voice to a high falsetto note. A student once described Reed's talent at speaking to a crowd: "As a teacher Dr. Reed always seemed to me to be first of all, master of his theme. His information was so much his own—a part of him, as it were—that when it was given to others it flowed forth with unadulterated naturalness, and sparkled with a keen interest which his charming personality could not help but lend it." That day, Reed's voice rose and fell as he outlined the objectives and results of the Yellow Fever Board. His voice grew higher as he emphatically explained that the control of the disease would be dependent upon the destruction of mosquitoes and protection against their bites. Two months of experiments in the Infected Clothing Building had not yielded one case of yellow fever. Reed had shown that "things"— clothes, trunks, linens, even corpses—could not transmit yellow fever. The tests conducted in the Infected Mosquito Building demonstrated that exposure to *Aedes aegypti* mosquitoes carrying yellow fever—and only exposure to those mosquitoes—spread the virus. And the blood inoculations verified that the virus was carried in human blood.

Reed ended his presentation expressing his regret for the fact that the discovery could only be made through human experimentation. He added that James Carroll, who was present that day, had been the first to succumb to the disease. Carroll received a standing ovation.

The applause roared in the auditorium when Reed finished his presentation, and he received handshakes from several Cuban, Spanish, Mexican, South American and North American physicians. The applause continued to echo in the walls of the lecture hall; it spilled into the hallways where people crowded to hear the monumental paper. It resounded like a heavy rainfall, and Reed listened to the gush of praise and congratulations. It was a sound that he had waited for his whole life, and he soaked it up.

It did not last long. Once again, the *Washington Post* seemed less than impressed with Dr. Walter Reed. In an article about the Pan-American Medical Conference, it begrudgingly accepted the board's mosquito theory, but asked, "Why not devote themselves to the eradication of the medium instead of killing more people by way of academic demonstration?"

CHAPTER 22

Retribution

On September 5, 1901, Walter Reed was given orders to proceed to Buffalo, New York, as the officer representing the Medical Department of the Army at the annual meeting of the American Public Health Association, where the majority of the discussion would focus on yellow fever. But the next two weeks would prove to be dark ones.

President William McKinley, reelected by even greater numbers thanks to the popularity of the Spanish-American War and Theodore Roosevelt as a running mate, took a trip through the West, ending in Buffalo, New York. His speech to the Pan-American Exposition on September 5 had attracted 50,000 people, and on the next day, September 6, 1901, McKinley made his way through the exposition crowds to the Temple of Music. In spite of the Secret Service, throngs of people approached the president, who patted men on the shoulder, shook hands and greeted bystanders. One, named Leon Czolgosz, pushed his way through the crowd. The Secret Service noticed a handkerchief in his hand, but by then, Czolgosz was within two feet of the president. McKinley smiled at the man and reached out his hand just in time to receive two revolver shots, one in his chest, and one in his abdomen. McKinley reeled from the shots, took a few steps backward and sat down in a chair before he was rushed to a hospital, and later, to the private home of John Milburn. Doctors operated on the president, but one week later, on September 14, he died from infection.

Two days later, on September 16, Walter Reed sat in the auditorium and listened to the president of the American Public Health Association begin the meeting with an honorary mention of both his friend Jesse Lazear and the engineer George Waring, who had invented the Memphis sewer system so many years ago and who died of yellow fever after a sanitation trip to Havana.

The doctors present that day were distracted by the death of President McKinley, but important matters were at hand, and focus quickly shifted to a discussion of Walter Reed's yellow fever study.

"I wish briefly to add my words of congratulation to those of the whole scientific world in praise of Dr. Reed and his colleagues, who, in my opinion, have given us a work which has not been equaled, as far as its benefits to the public health are concerned, since Jenner gave us vaccinia," gushed one doctor.

Still, there were those present who opposed the success of the Reed Commission, as it was becoming known. Eugene Wasdin, a longtime champion of Sanarelli bacteria, was also present. Having been on the board appointed by the president to confirm the Sanarelli bacteria in Havana, he had been incensed by the contradictory findings of Agramonte, then by the very public discovery of Reed's Yellow Fever Board. Wasdin must have felt strongly about the subject for he arrived from the bedside of McKinley; he had been the anesthesiologist and an attending physician for the president. Worse, rumors of dissension among McKinley's doctors now arose. Wasdin believed undoubtedly that the president's infection was the result of a poisoned bullet, and he said as much in a *New York Times* article. His theory would prove to be the wrong one. Amid this very public stress, he attended the annual meeting for the Public Health Association, primarily to dispute Walter Reed.

"The fact that Dr. Reed states that the organism has not yet been discovered does not make that true. The organism has been discovered, and it is not inconsistent with Dr. Reed's demonstration of the transmission of the disease by the mosquito, to accept the organism of Sanarelli as the cause of yellow fever . . . although Dr. Reed has demonstrated to my mind that the disease may thus be transmitted, it is not the only way by which we can contract the disease, and when contracted from the mosquito I deem it but an artificial infection such as we produce in animals in our laboratories . . ."

Reed rebuffed, "Will Dr. Wasdin please tell me in the transmission of malaria by the bite of the mosquito, whether that is also simply a laboratory disease?"

The familiar issue of quarantine battles ensued, especially among Wasdin and doctors from New Orleans, who had a hard time letting go of the theory that yellow fever could also be spread by infected clothing or bedding. With the New Orleans doctors, Reed was understanding. A city so threatened by yellow fever would approach any new study with some skepticism. For Wasdin, Reed had less patience.

"I was going to reach Dr. Wasdin's objections in a little while, but perhaps I had better answer him now ... It seems to me a waste of time, with all due respect to Dr. Wasdin, who has labored so hard over this problem, to longer consider this bacillus as the cause of yellow fever."

The meetings finally concluded with a resolution for President McKinley, "That the American Public Health Association has received with deep sorrow the intelligence of the sudden and tragic death of the beloved President of the United States ... in President McKinley we recognize the highest type of modern civilization, a patriotic citizen, a Christian gentleman, and a sagacious and enlightened statesman."

In the coming months, Wasdin began showing signs of mental illness and was eventually committed to an asylum where he died.

Reed's work also spawned new studies on yellow fever in Cuba. Dr. Juan Guitéras was attempting to produce immunity against yellow fever through mild cases. "He was probably the first medical authority to advance the theory that Cuban children were not born immune to yellow fever but acquired immunity by mild, unrecognized attacks in childhood," wrote Albert Truby.

Guitéras hoped that by infecting people with mild, nonfatal cases of yellow fever, they would be inoculated against it for life. As human cells fall under possession of a virus, the human body mounts a defense, building antibodies. Cells encounter a strange virus and its jagged edges in the bloodstream and create antibodies to attack the virus—the same way we shave and carve a metal key to fit a corresponding lock. But in this case, the lock of the antibody fits neatly against the key of the virus. As the human body struggles to create enough antibodies to lock onto and conquer

the virus, the illness takes its course. The host may survive or it
may die in the meantime, either from the symptoms of the virus
or the body's own immune response. In many viruses, one encounter with the disease produces a lifelong immunity. Those antibodies are always present, and if the person comes in contact
with the virus again, the antibodies render the virus useless before
it can do any damage.

Guitéras also believed that the success Walter Reed enjoyed
was the result of mild cases of the fever, which could be produced
by only one bite from a mosquito rather than a series of bites.
Reed didn't believe that to be the case and wrote to a friend: "I see
that Finlay and Guitéras continue to harp on the harmlessness of
a *single* mosquito's bite, drawing the conclusion that ordinarily
Y.F. is due to multiple *bites*. After some poor devil dies, they may
change their minds."

In what was known as "Guitéras Block" in the Las Animas
Hospital, Guitéras inoculated ten volunteers using the same pro-
cedures Reed had used at Camp Lazear. As usual, yellow fever
took on a life of its own. Guitéras's cases were not as lucky as
Reed's and three of the volunteers died. One, a young American
nurse named Clara Maass, died of yellow fever on August 24, exactly one year since Carroll and Lazear had started their ill-fated
experiments with mosquitoes. Maass had volunteered for the experiments in hopes that she would develop immunity, which
would enable her to be a more effective nurse. Though she was
bitten on four different occasions, the fifth proved fatal.

Reed wrote of their failure, "I was very, very sorry to hear of
Guitéras' bad luck and can appreciate fully his mental distress
over his loss of life—Perhaps, after all, the sacrifice of a few will
lead to the more effectual protection of the many."

While Guitéras failed to inoculate successfully against yellow
fever, he did inadvertently prove one thing: The virulency of one

case of yellow fever is determined by the virulency of another. In other words, strains of yellow fever vary in their deadliness. Clara Maass had been bitten a number of times in the months prior to August 1901. But that month, she and the other two fatal cases received blood from one source labeled Alvarez. The strain from patient Alvarez was apparently more virulent than others. That fact would come as little surprise to doctors who had served during Philadelphia, New Orleans and Memphis epidemics of yellow fever in decades past. Throughout history, yellow fever had swept through cities sometimes causing mild cases, other times killing thousands. For the virus, it was just a matter of fine-tuning.

One week before Clara Maass died, James Carroll returned to Havana to work with Guitéras on the experiments. This time, they would continue the search for the agent in the blood that caused yellow fever. Since Camp Lazear had closed, Carroll set up work in the plain, one-story Las Animas Hospital. Using the Berkefeld filter to catch any bacteria, Carroll passed blood samples through the filter, then injected volunteers. He produced cases of yellow fever. In doing so, Carroll had discovered that the agent that caused yellow fever was a filterable one. It was not bacteria that could be caught when filtered, nor was it a plasmodium, like malaria, that could be seen through the microscope. Though the term *virus* in its modern-day definition did not yet exist, that is exactly what he had found. James Carroll had isolated the first human virus.

Up to this point, a virus was an unknown entity. The actual word *virus* is Latin for venom, and that was the general definition. Science recognized that some poison was attacking the body; they just couldn't find it. Vigilant microbe hunters studied blood smears of ill patients looking for the germ that caused a particular illness. Bacteria of all shapes and sizes had been discovered and named. Malarial parasites had been seen. It made it hard to imagine something even smaller that could be even more deadly.

Following the American Public Health Association meeting in Buffalo, friends and colleagues began suggesting to Reed that he consider the job of surgeon general for the army, which would soon be vacated when Sternberg retired that year. Kean was chief among Reed's supporters. But the position would not go to Reed; his name was not even on the list of contenders. Too humble to actively seek such a position, Reed would complain in a private letter that he had too much common sense to consider himself in the race. He added, "The Moral of all of this is . . . to make friends wherever you go—political friends, if possible—Never mind about really accomplishing anything." But the bitterness was short-lived. In the same letter, Reed continued: "But, then, there is another way in which to look at this matter. Instead of simply being satisfied to make friends and draw your pay, it is worth doing your duty to the best of your ability, for duty's sake; and in doing that while the indolent stand you may accomplish something that will be of real benefit to humanity and worth more than all the high places that could be bestowed by shrewd politicians."

There was also talk that Reed might receive the new, prestigious award called the Nobel Prize; but, in 1902, it went to Sir Ronald Ross, who demonstrated the link between malaria and mosquitoes. Most people felt certain Reed would be awarded the prize in the future, but that was never to be the case. The Nobel Prize could not be given posthumously.

Strain began to show on Walter Reed; lines etched into his face and his hair grew ashen. He was only fifty-one years old, but seemed to be aging rapidly. Reed spent the summer of 1902 with his family in their summer home, Keewaydin, in Blue Ridge Summit, Pennsylvania. Keewaydin, which had been built by Reed, had windows and balconies that afforded views in every direction. The wooden shingle house had sprawling porches, white columns and a beautiful garden. It was a place he cherished with the inscription: *Love ever at my fireside, And peace within my door.*

A friend who visited him there that summer remarked, "He was very much worn by his scientific labours, but it was also evident that he felt most keenly the attempts which were being made by persons high in authority to rob him of his just fame for the work which he had done."

Reed had even written in a letter to his wife, shortly after his success in Cuba, predicting that Sternberg would attempt to take credit. "Of course," Reed wrote, "he will, at once, write an article and say that for 20 years he has considered the mosquito as the most probable cause of yellow fever. That would be just in order for him to do so." Reed was right. Sternberg published an article in *Popular Science Monthly* in which he claimed credit for the idea of the intermediary host. Reed wrote in a letter to a friend that "You might tell Dr. Finlay, too, with my best compliments, that he had better look to his laurels as the prosper of the Mosquito Theory, since Dr. Sternberg, in an article in the July *Popular Science Monthly*, puts forward his name very conspicuously for the credit for our work in Cuba." Reed added, "This is the reward for our work in Cuba! He knows, as well as I do, that he only mentioned Finlay's theory to *condemn* it!"

Sternberg could not bear to see another great medical discovery made without his name stamped upon it. He continually attempted to take credit for the discovery, and in 1905, when applying for a promotion in rank, Sternberg wrote: "I beg leave to call attention to the fact that the important discovery that yellow fever is transmitted by mosquitoes was due to my initiative. Without detracting in the least from the honor due Major Walter Reed and his assistants, who demonstrated this fact by a masterly series of experiments, the official records will show that this investigation was made upon my recommendation, and that the members of the Board were selected by me. I, also, gave personal instructions to the President of the Board, and pointed out to him the direction this experimental investigation should take."

Worst of all for Reed was the fact that his mind had grown weaker. As he was leaving his house for the Columbia lecture hall one evening, Reed shook his head and complained to his wife, Emilie, "I can't realize that I wrote this lecture, it is utterly beyond my mental capacity now."

By fall of that year, Walter Reed described himself as "a sick man." He returned home on November 12, in pain, telling Emilie that he must have eaten something disagreeable. His abdomen was tender, and he began to think it was appendicitis. Although he requested his favorite breakfast, waffles, Emilie took the advice of his doctors and refused him heavy foods. Reed spent the morning in bed reading the paper and planning the garden for Keewaydin. But that night, his temperature rose, and his friends, William Borden and Jefferson Randolph Kean, decided to operate. The next morning, he was sent to the Army Hospital at Washington Barracks, but even as he left his bedroom, he refused a stretcher and insisted on walking. He even stopped at his desk on the way out to write a check.

Major Borden, an expert on appendicitis, would perform the surgery, but Reed's longtime friend, Kean, would be there as well. In the operating room, an intern assembled the inhaler, a tank of nitrous oxide and an ether can. He asked Reed if he had any false teeth. "No," he said emphatically. As the ether began to pull at his consciousness, Reed turned to his friend Kean and said, "I am not afraid of the knife but if anything should happen, I am leaving my wife and daughter so little. So little, so little," he repeated.

The surgery, which took an hour and a half, revealed an enlarged and partially perforated appendix. It also showed signs of previous inflammation. Reed did not recover well from the operation and suffered from nausea and nervousness; his temperature stayed around 101 degrees, and his pulse began to rise, leveling out at 128. A junior medical officer was assigned each night to stay with Reed—Albert Truby was one of them—and Kean was there every day. In an attempt to lift his spirits, Kean told Reed that he would soon be promoted. "I care nothing for that now," replied Reed. "It was the reply of Lancelot to the Lake," Kean later said, *Prize me no prizes, for my prize is death.*

In spite of excellent medical care, the infection spread, peritonitis developed, and on November 23, 1902, Walter Reed died. All of the doctors present agreed that his health, weakened by the stress of his work in Cuba, had led to a fatal infection. Truby even believed that Reed unknowingly suffered from appendicitis during his work in Cuba, where Reed watched his diet vigilantly and suffered from stomach upsets.

It rained on Tuesday morning, November 25, when Reed was buried. The Potomac River blurred in the white haze, and the town houses of Dupont Circle and Georgetown cast tall shadows against the sidewalks. Clouds and rain cloaked the 100-foot spire of St. Thomas Episcopal Church. The Gothic stone building stood on Church Street, near Dupont Circle, and just a few blocks from the Reed's home. The church was full of military men, all in uniform, including Albert Truby. Truby, who had been given his first professional opportunity by Reed, would one day rise in

ranks to become brigadier general and the commander of the Walter Reed Army Institute. He watched as the guests filed in
through the wrought-iron and glass doors of the church trailed by
gusts of wind and golden leaves. Dr. William Welch, Reed's mentor and professor, was there, as was Dr. William Osler, Dr. Simon
Flexner and Secretary of War Elihu Root. William Randolph
Kean was one of his pallbearers. Emilie was so distraught, she
could not attend. And Reed's son, Lawrence, was stationed in the
Philippines at the time. Lawrence Reed received only a wire with
the news: *Your father died today.* It was two months before he heard
any more details; it would have cost additional money for the
army to include more in the message.

Following the service, Reed was buried in Arlington National
Cemetery, and his epitaph was taken from the recent remarks of
the president of Harvard University: "He gave to man control
over that dreadful scourge Yellow Fever."

Kean described his death as like the loss of a brother. And Dr.
William Welch said of Reed: "Doctor Reed's researches in yellow
fever are by far the most important contributions to science
which have ever come from any army surgeon. In my judgment
they are the most valuable contributions to medicine and public
hygiene which have ever been made in this country with the exception of the discovery of anesthesia."

Theodore Roosevelt, then president of the United States, remarked: "Major Reed's part in the experiments which resulted in
teaching us how to cope with yellow fever was such as to render
mankind his debtor, and this nation should in some proper fash-
ion bear witness to this fact."

Shortly after his death, a Walter Reed Memorial Association
was established to raise funds for Emilie and Blossom Reed, as

well as to finance a monument in Reed's honor. Contributions were made by Alexander Graham Bell, John D. Rockefeller, John P. Morgan, George M. Sternberg, as well as William Welch, Carlos Finlay and William Gorgas.

CHAPTER 23

The Mosquito

It would not take long for the mosquito theory to prove true. After the Yellow Fever Board left Cuba at the turn of the century, Major William C. Gorgas turned his focus to eradicating the *Aedes aegypti* mosquito in Havana, and later Panama, basing his work on Reed's findings. If science could not yet conquer the virus, man would at least destroy its accomplice.

Gorgas, a southerner who had been rejected from West Point, was known by a number of women as "the gorgeous doctor" around the army bases. He had been appointed to oversee sanitation in Havana during the Spanish-American War. He was also a close friend of Walter Reed's. The two men had a number of things in common. Both had studied at Bellevue; both had served as frontier doctors; both sought out knowledge, relishing books. But there were also differences. Reed was a physician who ended up in the military; Gorgas was a military man who ended up a doctor. Where Walter Reed would fight disease in the lab, Gorgas would attack it head-on.

Gorgas's campaign was one of the most successful and widespread sanitation campaigns in history. Not only adult mosquitoes but also their breeding grounds had to be destroyed. That meant addressing every single open-water source in the city, whether it was a flower vase, broken pot or puddle of water. Gutters were disinfected. Ponds were filled with larvae-eating fish. This was also a time when water was still stored in barrels and cisterns—perfect breeding grounds for the fresh-water-loving *Aedes aegypti*. Where the pipe connected to the barrel, Gorgas insisted screens had to cover the opening.

The city was divided into districts, and Gorgas sent his men into every home and dwelling in Havana for inspection. It was not an easy task. The locals would often hide containers when the "mosquito hunters" came through the neighborhood, but Gorgas would not be deterred. He soon had his officers keep accurate count of every container that could hold water—all had to be accounted for during an inspection. Barrels, jars and containers were given a slick, top layer of oil to suffocate mosquito eggs and wigglers. If people feared the oil or complained that it changed the taste of water, mosquito wire was strung across the top. Even mosquito traps were used—pans of fresh water were left out where the female mosquito could lay eggs. The water was then disinfected and the eggs killed before they could hatch.

Gorgas also attacked the problem at local hospitals. Fever patients were strictly quarantined with mosquito netting over their beds and strips of paper sealing any cracks in the wooden-frame buildings. Pyrethrum, an insect powder, was burned inside the room, and a light was held in the corner to attract wayward mosquitoes and stun them.

Finally, Gorgas made mosquito control a personal responsibility, sending out inspectors and fining citizens when mosquito larvae were found on their property or in their home. It is a practice still used today in Havana.

Gorgas's persistence proved highly effective. There was one yellow fever death in March 1901 and not a single one in the months of April, May or June. Malaria rates dropped dramatically as well. In their book *Yellow Jack*, John Pierce and Jim Writer wrote, "Yellow fever had been constantly present in Havana for 150 years and was nearly wiped out in less than 150 days." Gorgas himself would write: "It seems to me that yellow fever will entirely disappear within this generation, and that the next generation will look on yellow fever as an extinct disease having only a historic interest. They will look on the yellow fever parasites as we do on the three-toed horse—as an animal that existed in the past, without any possibility of reappearing on the earth at any future time."

Gorgas would go on to apply the same techniques in the development of the Panama Canal. French engineer Ferdinand de Lesseps, who had earned fame building the Suez Canal, had attempted to build the Panama Canal in 1881. The seventy-four-year-old engineer was met with disastrous results and lost as many as one-third of the men to yellow fever and malaria. Eventually, the project was abandoned. In 1904, Gorgas was assigned as the medical officer to America's Panama Canal project. President Theodore Roosevelt, a veteran of the Spanish-American War who could fully appreciate the devastating effects of yellow fever, fought to keep Gorgas in Panama in spite of political pressure to fire Gorgas and abandon his wild ideas about mosquitoes. Even Secretary of War William H. Taft pressured the president to remove Gorgas. Finally, a friend and doctor recommended to Roosevelt, "You must choose between the old method and the new;

you must choose between failure with mosquitoes or success without them."

Gorgas again applied his aggressive techniques toward destroying the mosquito and all of its breeding grounds. Once again, Gorgas met with success. Though he continued to receive criticism from skeptics during the first few years of the canal project, he outlasted them all; and he was still the officer on duty when, in 1914, the first ships sailed through the Panama Canal. Gorgas wiped out the mosquitoes, and the cases of yellow fever and malaria dropped off the charts. By 1908, William C. Gorgas had been appointed president of the American Medical Association, and then he was named surgeon general of the U.S. Army, occupying that office during World War I and the 1918 influenza pandemic. But what Gorgas most looked forward to was returning his focus to the eradication of yellow fever. In 1920, Gorgas traveled to London en route to Africa where he would take part in a yellow fever study. While in London, Gorgas was to be honored by King George V. Before he could attend his own ceremony, however, Gorgas suffered a stroke and was admitted to a London hospital. The king visited Gorgas there, granting him knighthood. Sir William C. Gorgas died four weeks later.

After the death of Juan Guitéras's volunteers in Havana, most notably Clara Maass, the public protest against human volunteers reached an all-time high. That, with the fact that the army had eliminated most of the mosquito's breeding places in Havana and surrounding areas, prompted the government to halt any further testing. Carroll returned to the United States. Though Sternberg recommended that Carroll be promoted to the rank of major, it was denied due to a moratorium on promotions. Instead, Carroll received a regular army commission as first lieutenant. It was not until 1907 that he would be promoted to the rank of major as a Special Act of Congress. In the same year, Carroll began suffering from heart problems. It was suspected that because of his age and the tremendous toll yellow fever took on his body, his heart had suffered irreparable damage. James Carroll died of valvular heart disease on September 16, 1907.

When the Yellow Fever Board was dismantled, Agramonte decided to stay in Cuba where he began teaching at the University of Havana. His work with yellow fever, however, did not come to an end. He continued to champion Dr. Carlos Finlay and also defended the commission's findings against future medical experiments involving yellow fever. Agramonte returned to the U.S. as a professor of tropical medicine at Louisiana State University in New Orleans, where he died in 1931. He was the only member of the Yellow Fever Board still alive to receive the gratitude of the U.S. government.

In 1929, through an Act of Congress, Walter Reed, Jesse Lazear, Aristides Agramonte, James Carroll and the men who volunteered for the human experiments were awarded the Congressional Gold Medal, so that their services, in the interest of humanity, may never be forgotten. The names of those men are: James H. Andrus, John R. Bullard, A. W. Covington, William H. Dean, Wallace W. Forbes, Levi E. Folk, Paul Hamann, James L. Hanberry, Warren G. Jernegan, John R. Kissinger, John J. Moran, William Olsen, Charles G. Sontag, Clyde L. West, R. P. Cooke, Thomas M. England, James Hildebrand and Edward Weatherwalks. Nearly thirty years later, two more names were added to that list: Gustaf E. Lambert and Roger P. Ames, the nurse and doctor who treated the majority of yellow fever cases at Camp Lazear.

* * *

Major William Borden, the friend who had performed Reed's operation, began a campaign to build what would become known as "Borden's Dream." Borden wanted to combine the Army Medical Museum, Army Medical School and Surgeon General's Library into a single medical center. And he wanted to name it for his friend Walter Reed.

It would take years to get the appropriate funding to buy the forty-three acres of land at Georgia Avenue, NW, and Sixteenth Street, NW, in Washington, D.C. The Walter Reed General Hospital would admit its first patients in 1909. Other buildings were added over the years, including the Army Medical School, which would one day become the Walter Reed Army Institute of Research. And the Surgeon General's Library, located downtown, would become the National Library of Medicine. The campus would change its name to the Walter Reed Army Medical Center in 1951 on the 100th anniversary of Reed's birth.

The redbrick hospital, complete with white columns and a fountain, originally held only 80 beds. When World War I broke out, that number jumped to 2,500. Today, the hospital admits close to 16,000 patients a year. Eventually, the center required more space and now has buildings in three different states, though the original hospital still stands in the District of Columbia. But not for long. The Walter Reed Army Medical Center is scheduled to close its doors for good by 2011, just over a century after it first opened. It will be combined with the Naval Medical Center in Bethesda, Maryland. Walter Reed's name will live on, however; the new campus will be called the Walter Reed National Military Medical Center.

PART FOUR

United States, Present Day

In recent years, popular attention has been drawn to . . . Ebola as the most frightening emerging infection of humankind. However, patients with yellow fever suffer as terrifying and untreatable a clinical disease, and yellow fever is responsible for 1000-fold more illness and death than Ebola.

—*Lancet Infectious Disease*, 2001

CHAPTER 24

Epidemic

It was March 10, 2002, when Tom McCullough checked into the emergency room in Corpus Christi, Texas. He had been suffering for four days from cramping, abdominal pain and severe headache. Then, he developed a fever approaching 103 degrees. The doctors in the ER thought it could be rickettsial disease, a term that covers a number of infections caused by vectors like ticks, fleas or contact with animals. Most rickettsia can be controlled with antibiotics, so the doctors prescribed just that and released him from the hospital. Two days later, he was back again, this time with intractable vomiting. McCullough had been a healthy, forty-seven-year-old man, but he now appeared weak and febrile. He repeatedly asked his wife, "What is happening to me?"

A series of tests were performed, and he was treated for malaria though his blood test proved negative. McCullough developed anemia, his blood would not clot, and his kidneys and liver failed. He went into shock and developed seizures. He bled uncontrollably from the sites of his needle punctures. Tom McCullough died on March 15—leaving a wife and six children still wondering why.

McCullough's illness and death were reported to the Centers for Disease Control and Prevention (CDC) in Atlanta, which began their own series of tests looking for dengue, St. Louis encephalitis, spotted fever, leptospirosis, Machupo virus and yellow fever—all viruses known to exist in South America. McCullough, it had been reported to the CDC, had just returned from a week-long fishing trip for peacock bass on Brazil's Rio Negro. The brochure for the trip read, *We do not suggest any inoculations of any kind for this trip . . . But to make sure you are worry free, consult with your personal physician.*

It would seem that some vicious new virus had taken hold of Tom McCullough; instead it was an ancient one. One hundred years ago, doctors would have known immediately what killed him, but modern medicine takes longer. Today, there is a wealth of illnesses known to be caused by insect vectors of all types. There are antibiotics and vaccines to fight disease, and still, this fever seemed to defy contemporary medicine. At last, the autopsy showed antibodies to the yellow fever virus—McCullough's internal struggle against a virus rapidly taking hold of his body. The CDC had reason to be concerned; McCullough was the third death from yellow fever since 1996, all three originating from trips to the Amazon region. Prior to that, there had not been a yellow fever death on American soil in nearly eighty years.

Tom McCullough had told his wife that he could not remember being bitten by a mosquito during the trip. He slept in an air-conditioned boat and had worn DEET. Still, a mosquito had apparently found him, following the scent of carbon dioxide in the tropical air, perhaps hovering unnoticed around his ankles or legs, biting several times as he moved. But it only took one bite, a pinprick he never even noticed, and the lethal virus made its way into his bloodstream. McCullough's body had never come in contact with this virus before. He had not had a yellow fever vaccine, and his blood came from stock that had not seen this virus in over a century.

Had an *Aedes aegypti* mosquito in Texas bitten McCullough in the days before he checked into the hospital, hundreds more could have been infected. The virus would have been unleashed on a virgin population. In the mild Corpus Christi winter, virulent eggs could survive to the next summer when even more *Aedes aegypti* mosquitoes would carry the virus through another muggy Texas summer.

At first, the virus would move quietly into the population. People would begin showing up at local emergency rooms with high fevers and flu-like symptoms. They would be released when they showed signs of improvement—yellow fever's convalescent period. But as many as 50 percent of those people, and possibly many more than that, would enter the toxic phase of the disease and die. Their deaths might be blamed on any number of diseases—pneumonia, hepatitis, influenza, West Nile. Though mosquito bites, swollen and pink, might appear on the skin, no one would think to investigate further. After all, these patients live in the United States. They had not traveled to a tropical country; they had just spent a summer evening outdoors, or found a striped mosquito trapped in their car, or missed a few places of skin when they sprayed Off! on their children playing in the backyard.

As the death toll began to mount, doctors in the local hospitals would begin reporting them to the state health department. Perhaps malaria or dengue had made its way from Central America north. Health officials would be concerned. Resistant strains of malaria have been reported in recent years, and the CDC estimates that as many as 3,800 cases of dengue have appeared in the United States since the 1970s. Dengue is spread by the same mosquito that carries yellow fever. At last, the dead arriving from their homes or on gurneys in emergency rooms would begin to yellow, their skin taking on a bronze color, their eyes like sunflowers.

The state health department would contact the CDC, which, under international law, must contact the World Health Organization within twenty-four hours to report any disease with jaundice and bleeding. Since its inception in the 1950s, the WHO's International Health Regulations have required reporting of only three diseases: plague, cholera and yellow fever. All three diseases are subject to international quarantine.

But in America, these diseases are so rare that doctors would doubtfully even recognize the symptoms in twenty-four hours. Americans traveling to the coastal areas of Texas for vacation would pick up the virus and fly home to cities like Houston, Dallas, Memphis and New Orleans, where entire colonies of *Aedes aegypti* live.

In 2005, the CDC published a detailed response to an epidemic of yellow fever in Africa and the Americas. Field investigators, border officials and vector control would arrive. They would contact the Global Alliance for Vaccines and Immunization to report an epidemic and request that mass vaccines be delivered within the week. Those who already have the virus would have little chance for survival—they would be part of the nonimmune population, the kindling that the virus relies upon to spread. Vaccines would be given to hospital personnel and military first, but postexposure, it would do little good. In the time it would take the vaccine to prompt the production of antibodies, the virus would have run its course, leaving its host either immune or dead.

A live vaccine, yellow fever can also have adverse effects. Infants, patients with depressed immune systems or anyone over the age of seventy-five cannot receive the vaccine. Though pregnant women are usually denied the attenuated vaccine for the safety of the fetus, the CDC would make an exception in the case of an epidemic. In the hospitals where yellow fever patients arrive, rooms would have to be screened and strictly quarantined. Lab technicians handling blood samples would have to follow strict procedure with gloves, masks and air purifiers.

A general panic would settle into the city and surrounding ones as educational warnings on television and radio recommended that people cover their beds in netting. Informational pamphlets would instruct people to empty any outdoor water containers around their homes. In spite of the summer heat, people would wear pants, long sleeves and socks with shoes. Store shelves would be cleared of Off! and any other DEET products. Windows would be screened. Water and food stockpiling might occur as people prepared to board themselves up in their homes, keeping their children indoors. Public pools and parks might close. Chemicals would be pungent in the air as people sprayed insecticides on their lawns and in their homes. Vector control units would send out patrols of trucks and crop dusters to mass spray.

The panic would worsen.

Vaccines from the Global Alliance for Vaccines and Immunization would arrive, but not enough in the event of a full-scale outbreak. The GAVI only recently began stockpiling the yellow fever vaccine. Six million doses are reserved each year for an epidemic, and they could take a few million more from their reserves for routine vaccine usage. The CDC would assess which portions of the population are most in need of the vaccine, reserving several for the personnel, military and hospital staff. Even if all six million vaccines arrived in a town like Corpus Christi, there would not be enough to inoculate cities the size of Houston and Dallas, much less other southern cities where the mosquitoes or infected people may have made their way.

Cases would continue to appear well into December, spiking every time another warm front moves through the country. At long last the epidemic would subside, though it would live on in the news and on the covers of magazines for months. Major vaccine production programs would begin, grown in chicken eggs over the next six months. And, hopefully, there would be enough vaccines ready for the approach of warm weather the following spring when yellow fever season arrived once again. That is not always the case—especially in underdeveloped countries. After an outbreak of yellow fever that killed thousands in Nigeria during the 1990s, it took ten years to clear the population of the virus. In order to prevent an epidemic, at least 80 percent of a country must have immunity to yellow fever.

According to the World Health Organization, even a single case of yellow fever must be treated as epidemic.

CHAPTER 25

A Return to Africa

Dr. Adrian Stokes bound a monkey onto a cushioned board with gauze, keeping his head firmly strapped. For an hour, Stokes allowed *Aedes aegypti* mosquitoes to bite the monkey on his face, lips, ears. Then, he returned the monkey to its cage. It seemed a little cruel, but it was too dangerous for the doctors to hold the monkeys while loaded mosquitoes fed. Even with leather gloves on, the insects could bite through the stitching. Across the lab from the monkeys, in a cage with roughly six screens dividing it, mosquitoes hummed in their wire prison.

A forty-year-old doctor who worked in pathology at Guy's Hospital Medical School in London, Stokes was a part of the Yellow Fever Commission sent to West Africa in 1920—the one William C. Gorgas was to be a part of when he died in London. Stokes was a graying Englishman—charming, a tennis player, loved by all those who worked with him. Stokes rarely wore gloves when he worked and took poor care of his hands and fingernails. One doctor had even noticed an open sore on Stokes's finger from a monkey bite.

The commission's lab was located in Yaba, not far from Lagos, Nigeria. Africa at the start of the twentieth century was still very much the "Dark Continent"—dark primarily because of colonial Europe's inability to understand it. For centuries Europeans and Americans had landed on the coast of Africa to enslave Africans; now, they landed on the same shores to enlighten them. The white man, it seemed, would not stay out of Africa; but Africa would not have him. In The Coming Plague, Laurie Garrett wrote of the song African children sing championing yellow fever: "Only mosquito can save Nigeria, Only mosquito can save South Africa, Only mosquito can save Zimbabwe, Only Mosquito can save Africa, Only malaria can save Africa, Only yellow fever can save Africa."

By the 1920s, after Reed's proof of the mosquito vector and the vigorous campaign by Gorgas to eliminate yellow fever in urban environments, yellow fever, it seemed, would soon be conquered. Paul De Kruif, in his popular book, Microbe Hunters, wrote: "Because, in 1926, there is hardly enough of the poison of yellow fever left in the world to put on the points of six pins; in a few years there may not be a single speck of that virus left on earth." Hubris. Arrogance. The lessons of the previous century diminished in the distance as the wheels of the Progressive Era rolled forward.

The same year that De Kruif wrote that statement, 1,000 people in an African village of only 5,000 became infected with yellow fever. Although the fever routinely hit nonimmune populations like British and French colonials, the colonial doctors had never seen an epidemic among native Africans before. It was as though an ember had been left smoldering in the jungle, and now fires were beginning to erupt.

America had entered the Progressive Era. Gone were Victorian ideals, heavy-laden tradition, elaborate ceremony and sentimentality. This was the age to move forward. Everything from home life to education to medicine took on a rational precision. For the first time in American history, more people lived in urban environments than on farms. Ideas, people and governments were compartmentalized—there were experts to lead the masses. Medicine had finally become a profession, and science was at the center of progressive thought. Greer Williams wrote that the old bacteriologist was not dead; he had merely "shaved his beard, put on horn-rimmed glasses, changed hats, and reappeared in a new branch of microbiology as a virologist."

The Rockefeller Institute embodied all the ideas of the Progressive Era. Johns Hopkins had been the premier place for the study of medicine since the 1870s, and it had made monumental discoveries, much like the European institutions before it. They lifted the veil and unraveled the mystery around disease, adding name after name to the list of famous scientists. The Rockefeller Institute, which opened its first lab in 1904, would take a more aggressive tack. It would not model itself after European research and individual scientists, but instead make a departure from it, moving beyond the study of disease to the eradication of it. Like the Progressive Era itself, the institute would take the offensive, its mission "to promote the well-being of mankind throughout the world," its vision "to cure evils at their source." Over thirty years, the Rockefeller Foundation would spend fourteen million dollars in an effort to eradicate yellow fever. It would seek out the fever and control the natural world. And there was no place where the natural world was still so unconquered as in Africa, no place where the well-being of mankind was more at stake.

A large cast of American and British doctors was sent to Nigeria by various organizations, including the Rockefeller Foundation, to eradicate yellow fever. The Rockefeller team needed a good pathologist, and in 1927, Adrian Stokes was assigned to work with them. They were the plague hunters, and their aim was not only to see that yellow fever disappeared from North America but to destroy it entirely.

In spite of bubonic plague, malaria and yellow fever, the scientists from both countries made the most of their time in the tropics. They played croquet on the lawn and dined together. They kept gardens with orange bursts of marigold and salmon-colored hibiscus and sun-streaked zinnias. They visited nearby villages in the touring Dodge. But there were a few cultural differences. The Englishmen usually stopped working after 4:00 in the afternoon to enjoy cocktails or play golf or tennis. The British, in turn, found the Americans and their penchant for sunshine strange. The Americans were forever bareheaded in the sun or screening their bungalows and blocking the breeze.

A new epidemic spreading among Africans gave Stokes and his partners fresh fever cases to work with. They drew blood from sick humans, storing it in a glass forest of vials and petri dishes in the lab, and injecting it into various lab animals, primarily monkeys. Their conclusions were mixed, and shipments of fresh monkeys and guinea pigs continued to arrive by boat and train at the lab.

The doctors were beginning to lose hope when they heard of a new outbreak of fever in Kpeve, Gold Coast—now Ghana—at the end of June. Much like Walter Reed and his Yellow Fever

Commission had done in Cuba, the doctors hunted down the disease wherever outbreaks of fever occurred. Members of the commission traveled the 100 or so miles from Accra to Kpeve to see a European farmer and his wife—both of whom had been diagnosed with typhoid. Instead, they found yellow fever; or perhaps, yellow fever found them.

They also discovered an African man, twenty-eight years old, named Asibi. He was sitting on a stool, his head in his hands, his temperature 103 degrees. *Aedes aegypti* mosquitoes swarmed around him like sparks rising from a flame. The doctors took blood samples from several of the patients, including Asibi, and returned to their lab in Accra.

Asibi's blood was injected into a marmoset, two guinea pigs and a monkey with the affectionate name Rhesus 253-A. The rhesus monkey arrived in a shipment from Asia; the doctors had discovered that African Old World monkeys seemed immune to the fever. Rhesus 253-A was the color of sand, with round, black eyes and a face like a human child. It was a lively, chattering monkey until it became ill a few days later. It grew quiet in its cage and soon died. Stokes autopsied the animal and found all the postmortem signs of yellow fever. It was their first major breakthrough. The doctors bled the monkey and injected the loaded blood into another monkey, Rhesus 253-B, and it too soon died of yellow fever. They passed the infected blood through a Berkefeld filter, and just like James Carroll, found nothing. No known bacteria or parasite had been caught in the filter. Whatever organism infected the blood of the monkeys had to be even smaller.

Mosquitoes fed on the blood of the sick monkeys, then the doctors bound healthy animals to boards, allowing the mosquitoes to bite, passing the virus from one monkey to the next, creating in the lab what nature had been accomplishing for centuries. If yellow fever could be passed effectively to monkeys, the possibilities were endless. Suddenly, the disease began to make more sense. Forest workers often returned from the bush with a case of yellow fever, in spite of the fact that they had not been in contact with any sick humans. If monkeys could harbor the virus, then the jungle itself was fueling the yellow fever virus, giving it refuge. It explained how endemic yellow fever lives quietly in the jungles, moving through monkey populations before exploding on urban, human ones. By 1935, an American doctor named Fred L. Soper would discover that monkeys also acted as hosts to the disease and that other mosquitoes could carry the virus as well. It became known as jungle yellow fever.

The discovery that the West Africa team had found a filterable virus passing through monkeys was met with controversy—particularly from a bacteriologist named Dr. Hideyo Noguchi. Noguchi, a doctor at the Rockefeller Institute with celebrity status in medical circles, believed he had found the *spirochete,* a rod-shaped bacteria, that spread yellow fever. What Noguchi lacked in physical stature, he made up for in intelligence and arrogance. He had been born to a Japanese servant family and changed his name to Hideyo, which means "great man of the world." It became his personal mantra.

Noguchi was famous for working beneath the Rockefeller Institute's director Simon Flexner. Most viewed Noguchi as either insane or a genius. He was well liked, but also egocentric, rabidly ambitious and a loner in medical research, preferring to conduct his experiments alone, and thus taking sole credit for them as well. His findings on yellow fever had been bold, but unsubstantiated and blasted by the likes of Aristides Agramonte—the only survivor from Walter Reed's original team and now an old man. When Noguchi heard of the latest discovery in West Africa—one that discounted his bacteria—he decided to set up his own study there.

As Noguchi readied himself to travel to Africa that September, he was met with word that Adrian Stokes was ill with yellow fever. The doctors guessed that Stokes's open wound on his hand had been infected with yellow fever blood. It was the first case of yellow fever transmission through skin. Stokes had retreated to his bungalow one night during dinner, where he began vomiting. Shortly thereafter, he moved into a hospital in Lagos, Nigeria. Ever the scientist, Stokes decided to personally take part in the experiments he had been performing on his monkeys. He insisted that his partners allow mosquitoes to bite him—200 in all. The doctors also drew blood from Stokes.

Two days later, on September 17, Stokes felt better. He read books and spoke of going to his lab to work, but he also made his colleagues promise that if he should die, they would autopsy him for the study. Stokes's temperature stayed around 101 degrees, while his pulse hovered at 70. His mind began to slip, at first just growing dull and later delirious. His skin grew yellow. Stokes died of yellow fever on September 19.

His colleagues, Bauer and Hudson, deliberated what to do next. They had promised to autopsy Stokes, but carving into their own friend an hour after his death seemed impossible. Bauer, with tears in his eyes, said he would do it. But Hudson, who performed all autopsies for the team, stopped him. "I will do it."

The diagnosis was a clear case of yellow fever, and the doctors had shown that the virus could cause an infection through the skin—their friend and colleague, Adrian Stokes, had been proof of that.

* * *

Two months after Stokes's death, on November 17, 1927, Hideyo
Noguchi arrived in Accra to begin his work. Since there had been
no new outbreaks of yellow fever, he would have to use blood
samples from the Asibi strain and Stokes. Again working alone—
and demanding a larger bungalow for his personal quarters—
Noguchi went to work proving his bacteria theory, or at least
proving that there might be different types of yellow fever. He
used roughly 1,200 monkeys during his six-month stay, spending
close to $20,000 on his subjects. In cables to New York, which
could run $1,800 per month, Noguchi declared, "My work is so
revolutionary that it is going to upset all our old ideas of yellow
fever."

By May, Noguchi had plans to return to the United States with
his proof in tow. The other doctors found his data confusing and
inconclusive, but he was not deterred by the opinions of less brilliant men. First, however, he wanted to visit the lab in Lagos to
compare studies. As he boarded the boat, Noguchi complained of
a chill. He looked tired and asked Hudson to draw some blood to
test for malaria. No malarial parasites could be found. Noguchi
made the overnight boat trip back to Accra, growing more ill. Another doctor housed Noguchi and cared for him during an illness
that was looking more and more like yellow fever. As his temperature rose, and his pulse slowed, he began retching the telltale
black vomit. He suffered from seizures and bit his tongue. Finally,
on May 20, his kidneys failed, and he started convulsing. Noguchi
died of an unmistakable case of yellow fever. The head of the West
Africa commission, Dr. Beeuwkes, and a British doctor named
William Young visited Noguchi's lab while he was in the hospital.
They found several monkeys dead in their cages; they killed the

rest for safety. They also found *Aedes aeygpti* mosquitoes flying around the room; the insects had somehow freed themselves from their corner cage.

Ten days after Noguchi's death, William Young died of yellow fever, presumably from the mosquitoes in Noguchi's lab.

Yellow fever would also kill another famous physician. Dr. Paul Lewis was a quiet, brilliant doctor who had discovered that a virus was responsible for polio and had been one of the main doctors in the fight against the 1918 influenza epidemic. Lewis, on assignment for Simon Flexner, would die of yellow fever in a tropical lab in Brazil.

And, finally, the virus would take its last victim: Theodore B. Hayne. Hayne was a researcher working for the Rockefeller Foundation in 1930 when he was sent to Lagos, Nigeria. He was thirty-two years old when he died there.

Yellow fever had always been and always would be a disease that countered every gain with a substantial loss.

It seems only natural that a virus should fight for its own survival, and yellow fever had been hunting down and killing the scientists attempting to destroy it. First, the fever killed Jesse Lazear outright, and later, James Carroll from presumed complications. It had indirectly weakened the health of Walter Reed, who died within two years of his study linking yellow fever to mosquitoes.

When the plague hunters turned their attention to West Africa, the virus's natural habitat, it struck again with just as much violence. Adrian Stokes, Hideyo Noguchi, William Young and Theodore B. Hayne died of yellow fever, even as they studied ways to control it. Then, Paul Lewis died in Brazil. During the Rockefeller Foundation's fight to eradicate yellow fever, five doctors in

all died of the fever, and a total of thirty-two scientists and technicians would contract the fever in the lab.

One doctor would later write: "I can think of no other disease
that killed so many scientists studying it."

CHAPTER 26

The Vaccine

The yellow fever virus, from its earliest beginnings in the African forests and savannah, to its widespread epidemics on the other side of the world, to the hundreds of thousands of its victims, has been connected by one thing: blood. It is blood in monkeys that harbors the virus. It is blood that passes the virus into the body of a mosquito. It is blood that connected 5,000 deaths in Memphis to Walter Reed's human experiments twenty years later. Blood is the medium that allows the virus to travel distance and time, passing from species to species. And it was in the blood that science finally found a way to fight the virus.

Max Theiler did not look like a man who would achieve greatness. He did not have the English charm and graciousness of Adrian Stokes. He did not have the larger-than-life self-confidence of

Hideyo Noguchi. For one thing, Theiler was five two. For another, he did not have a stellar academic record. Theiler had been born to Swiss parents in South Africa; he was schooled in London and lived in New York. Theiler never actually received a degree as a doctor in medicine or science, in spite of attending courses at the Royal College of Physicians and the London School of Tropical Medicine and Hygiene. When colleagues mistakenly called him "doctor," he never bothered to correct them, not because it embarrassed him, but because it didn't. "You can't educate a person; you can only create an environment in which he can educate himself," Theiler once said of his background.

Theiler was young, shy and kept to himself. His lab was littered with ashtrays and boxes of Chesterfields. He loved to read, enjoyed art galleries and found no interest in practicing medicine because there was too little to be done for the patient. In 1922, when asked if he would like to join the staff of Harvard Medical School, Theiler said, "Sure, fine."

Though many doctors were still searching for bacteria in the blood that could be linked to yellow fever, Theiler began to have ideas of his own. For decades a long list of skilled bacteriologists that included names like Sternberg, Sanarelli and more recently Noguchi had been searching for the *bacteria* that caused yellow fever—this, in spite of the fact that James Carroll had shown that yellow fever blood passes through a filter without catching any known bacteria. Theiler began to wonder if there wasn't something even smaller, something that could pass through a filter, something incapable of living in dead cells.

During the summer of 1929, Max Theiler's boss at the lab in Boston went on vacation, and Theiler decided to try his own experiment. Since monkeys were costly at fifteen dollars each, he opted for mice, which cost only a few cents. First, Theiler injected bits of yellow fever–tainted liver into the brains of mice.

They did not develop yellow fever, but died nonetheless, developing a sort of encephalitis. Next, he tried injecting the same blood
into the abdominal cavity of the mice—they lived. Each time, he
drew new blood from the mouse. He purchased three rhesus
monkeys and injected each with the mice blood. The first monkey died of yellow fever, the second developed a fever but survived, the third developed nothing at all. The yellow fever virus, it
seemed, had been turned upside down and inside out—it was
killing mice that were not supposed to be able to contract yellow
fever, and it was *not* killing the monkeys that were highly suscepti-
ble to it. In Theiler's hands, the virus could become more deadly
in one animal and less so in another. What Theiler had was a vaccine in the making.

    Theiler cannot be credited with this line of thinking. After all,
Edward Jenner is considered the first to have achieved this with
cowpox at the end of the eighteenth century. Jenner named the process *vaccine* from the Latin word for cow—*vacca*. But the basic
premise was the same: a virus can be manipulated, taking something
harmful, and creating out of it something protective. As the virus is
passed into another animal it adapts to its new host, it mutates into
a less harmful form. The scientist becomes God and the virus his
subject. The danger comes in the fact that a virus, ever mindful of
evolution and its survival, has its own methods of defense. If man
can manipulate the virus, the virus can manipulate man.

    Aside from its brute force in taking over cells, the other major
weapon in a virus's assault on the body is the ability to mutate.
Complex creatures, like humans, store their genetic material in
DNA, which is more stable and less likely to change. Viruses are
often made of RNA, an unstable store of genetic material, which
can produce errors when it replicates known as mutations. The
mutations can work against the virus, hindering it or even killing
it; in other cases, they enable the virus to kill more efficiently.

HIV is a prime example of a virus's skill at changing forms and mutating easily. Just when the body produces the right antibodies to fight a viral strain, the virus alters its outer coating just slightly. The key will no longer fit into the lock. That is why a flu vaccine, made up of several different influenza strains, is a yearly vaccine and not one that produces lifelong immunity. That is also why an influenza pandemic would prove so deadly. In the six months it would take to isolate the virus, grow it in chicken eggs and create the vaccine, the virus may have spread through the population decimating millions of people.

In spite of carrying a single, simple strand of RNA, yellow fever does not mutate easily. Instead, flaviviruses like yellow fever somehow disable the body's immune response—a process that continues to elude science. When the body encounters the virus, it mounts a mass campaign against the foreign invader on two fronts. White blood cells known as B cells create antibodies that cleave to the virus and mark the virus for destruction. At the same time, Killer T cells search out infected cells and destroy them. During the several days it takes to organize and implement this counterattack, the virus courses through the body, taking over kidney cells, liver cells, breaking down blood vessels until the organs themselves fail, and blood flows uncontrollably. There is no way to stop it unless the body can mobilize its forces before the virus has taken hold—that is what a vaccine does. The vaccine is a milder but live form of yellow fever that activates both arms of immune response—in advance.

Max Theiler's work in Boston caught the attention of some scientists working with the Rockefeller Foundation—one of which was Dr. Wilbur Sawyer. The Rockefeller Foundation offered Theiler double his salary at Harvard to join their yellow fever lab. Theiler accepted and began work with Sawyer on a vaccine to inoculate the doctors who worked in labs—doctors who continued to die from their work with yellow fever. Theiler himself had contracted yellow fever in the lab in 1929, but recovered, developing immunity. Recently, three other doctors had also developed yellow fever in their work in the United States and in Lagos; one had died. A vaccine would finally give the scientists trying to conquer yellow fever a fighting chance.

Sawyer and Theiler developed a makeshift vaccine combining the infected mouse-brain tissue with blood from a human immune to yellow fever. Then, they took the amalgam of virus and antibodies and injected it into a man named Bruce Wilson. Wilson, who had earned fame as a public health field director fighting against malaria, had just returned from Brazil. He was checked into a screened room at the Rockefeller Institute, where he was injected with this new vaccine. His temperature and pulse were monitored constantly, and to pass the time, Wilson taught his night nurse how to play poker. Wilson never grew ill, and instead, developed immunity to yellow fever.

They now decided to turn their attention to developing a large-scale, safe vaccine. French studies using Theiler's mouse strain had produced some negative reactions. Even Sawyer had seen the occasional case of fatal encephalitis during his test studies on monkeys. They decided to try an entirely new vaccine using milder strains of yellow fever taken directly from monkeys, forgetting the mice all together. An extensive laboratory was set up in which thousands of flasks, a factory line of glass tubes, housed the virus—part of the Asibi strain—and reproduced it in various forms. They experimented with mouse embryos, then chicken embryos. With time, they developed what became known as the 17-D vaccine, grown in a chicken embryo and named for the seventeenth series of experiments and the type of tissue used. The only complication wasn't really a complication at all: The virus required a little human nonimmune blood, a serum, to survive. The doctors therefore added about 10 percent human blood to the vaccine. It was cheap and safe—it seemed simple enough. But man continually underestimates nature, and as a result, nature occasionally makes folly out of man's triumphs.

Scientists had now been working on a vaccine for nearly ten years. Throughout the 1930s, as America fell into a deep depression, the doctors locked themselves away in labs and engaged in this viral fertility study. They nurtured the yellow fever virus, disciplined it, fed it blood and grew it in the surrogate confines of a chicken egg.

By 1941, it seemed clear that America would soon be at war, and as every war in the past had taught, disease could be far more devastating than the enemy. Dr. Wilbur Sawyer worked on a mass production of the vaccine to inoculate American soldiers as they left to fight. Theiler had some reservations about the new vaccine though. He wanted to try a serum-free vaccine, one that could be used without introducing human blood into the mix. Sawyer thought it was a good idea in theory, but added that there simply wasn't time. America was headed for war, and soldiers were headed into countries where yellow fever was rampant—particularly Africa. "You are courting disaster," Theiler told him.

In the fall of 1941, the yellow fever vaccine was given to all troops departing for the tropics, and by 1942 seven million doses had been issued by the International Health Division to the U.S. Army, Navy and the British fighting in Africa. But, complications began to arise—infectious hepatitis was reported among soldiers. At first, it was sporadic; then, it became epidemic. The soldiers seemed jaundiced, complained of headaches, nausea and dizziness. There were a few fatalities. It did not take long for the blame to fall on the new yellow fever vaccine—the blood that fed the virus in the vaccine had been taken from several hundred volunteers. A few of those—maybe 2 percent—reported a history of jaundice. Their blood had been pooled, and roughly 400,000 doses of the vaccine had been tainted. It became known as "Rockefeller disease" and "serum hepatitis."

In the end, there were close to 50,000 cases and 84 deaths. But there was not a single yellow fever case in an American soldier. Dr. Wilbur Sawyer took complete responsibility for the hepatitis epidemic, as well as for the notable absence of yellow fever among the troops. He would be remembered for the former, not the latter.

It was October 15, 1951, when Max Theiler received a cablegram at his lab in New York. He had been awarded the Nobel Prize in Medicine for his "discoveries concerning yellow fever and how to combat it." When asked what he would do with his $32,000 prize money, Theiler responded: "Buy a case of Scotch and watch the Dodgers."

Max Theiler is the only scientist ever to receive the Nobel Prize in connection with yellow fever.

CHAPTER 27

History Repeats Itself

Today, *Aedes aegypti,* the striped house mosquito, is blamed for any urban outbreak of yellow fever, but several other mosquitoes are known to carry the yellow fever virus as well. Those mosquitoes play a part in what's known as jungle yellow fever. They live in the tree canopies of Africa and South America and pass the virus back and forth through wild monkeys. When a human becomes infected it is because he, and they are usually young men, is working in the jungle clearing forests. In fact, jungle yellow fever is considered more of an occupational hazard than anything else. It survives on a continual cycle between mosquito and monkey, with the occasional human getting caught in the crossfire.

An urban epidemic of yellow fever occurs when jungle yellow fever makes the jump into a large human population. The forest-dwelling mosquitoes, perhaps an *Aedes africanus* in Africa or a

Haemagogus in South America, fly beyond the borders of the jungle into the territory of a city-dwelling *Aedes aegypti*. The two mosquitoes share a blood meal from a monkey, and suddenly, the virus is passed from a jungle mosquito to a city mosquito that spreads the virus to a human population. In Africa, the most common type of yellow fever is the intermediate one, in which yellow fever can survive in terrains between the jungle and urban cities. Whether it starts as a jungle outbreak in South America or an intermediate one on the African savannah, the worst-case scenario is the same: The fever moves into a large city, the virus builds more strength, and it infects thousands. An outbreak of urban yellow fever is always considered an epidemic.

Because the virus is not part of an urban cycle the way it is in the jungle, the virus is unleashed on a fresh population of nonimmunes. It is much the same as it was 400 years ago when Europeans first arrived in Africa during the slave trade. They landed on the shores of West Africa armed and ready to export human labor; instead, many served as nothing more than an import of nonimmunes for the yellow fever virus.

In just the same way, the cycle happened on this side of the world. It attacked nonimmune populations that had never before seen the virus. *Aedes aegypti* mosquitoes traveled on board the ships from Africa and then proliferated, most likely spreading yellow fever to native, forest-dwelling mosquitoes in South and Central America that settled back into the jungles to begin the cycle of mosquito and monkey transmission, harboring the fever in the sultry haunt of lush tropical life.

In the United States, the cycle had taken a different turn. There were no cases of jungle yellow fever, no forest-dwelling monkeys giving refuge to the virus; it was not a yearly occurrence. Instead, there was only a series of urban epidemics when the virus exploded on a population.

The cycle would not be broken until the mosquito's breeding places were destroyed. In Memphis and elsewhere, it happened through the invention of the sewer system and elimination of private cisterns and privies.

Massive campaigns against *Aedes aegypti* essentially wiped out the mosquito from Central and South America, and the U.S. government promised to do the same. In their book *Mosquito*, Andrew Spielman and Michael D'Antonio wrote that there was a sense of irony to the situation: "After all, America had gone to war with Spain, in part, because of the danger of yellow fever spreading from Cuba into nearby lands." Now that Latin America had the same concerns about the United States, there was opposition to the idea. North Americans did not welcome the intrusion of government employees trampling through yards and hunting mosquito larvae. What's more, the Environmental Protection Agency banned the use of DDT. As a result, *Aedes aegypti* never fully left the United States; if its presence today was not already known, it was brought to our attention with recent outbreaks of dengue in Texas.

Over time, mosquitoes have proven their evolutionary dexterity, adapting to insecticides and building a resistance. Once again, the striped house mosquito now flourishes in cities throughout South America, Central America and the southern United States. Its lyre-marked body and striped legs swarm around potted plants, gutters and rain-filled watering cans. A homebody, the domestic *aegypti* prefers human habitations, houses, boats and fresh water. But this was not always the case.

Hundreds of years ago, *Aedes aegypti* lived only in the jungles of Africa, where it hovered around tree trunks to lay its eggs in pools of rainwater. Its range was short; the mosquito preferred to stay in one general place, close to the trees. As man traveled into the interior of Africa, the mosquito made its evolutionary leap: It adapted to human life. Rather than tree trunks, it first sought water casks, then standing water around homes, and in modern times, oddly enough, it has adapted to tires. The dark interior and the water that clings to the inside of a tire mimic the hollows of a tree trunk. Scientists believe that the world's mass of discarded tires in urban settings has recreated the atmosphere of those ancient jungles in Africa.

Aedes aegypti is not the only mosquito to have made the journey from tree holes to tires. The Asian tiger mosquito, *Aedes albopictus*, is relatively new to the United States. The tiger mosquito arrived in 1983 in a shipment of tires from Asia. The large, striped mosquito is very similar in appearance to its *aegypti* cousin, but true to its name, the tiger mosquito is a more voracious feeder. As a vector the tiger mosquito has been known to carry dengue, encephalitis and yellow fever in other countries. It has yet to transmit disease in the United States. The tiger mosquito, hardy and determined, has proliferated in North America, even crowding out some of its *aegypti* neighbors. Nature has a dark sense of humor though: The first Asian tiger mosquito on this continent was found hovering in Elmwood cemetery in Memphis, Tennessee.

In recent years, vaccine usage for yellow fever has fallen off. The problem is a lack of education and funds sufficient for the dissemination of the vaccine. Surveillance is also minimal. As a result, in the 1970s, outbreaks of yellow fever began once again, creating what is known as the "yellow fever belt" in Africa. The World

Health Organization described the 1980s and 1990s as an extraordinarily active period for yellow fever in Africa. But that activity continues even today, and Nigeria has reported more epidemics than any other African country.

Ninety-three percent of the countries in West Africa now have cases of yellow fever—up 30 percent just since the mid-1990s. Yellow fever, it seems, is making a comeback; it is also spreading to areas that have never before seen the virus. Thirty-three countries in Africa and nine in South America are now known to house the virus. The number of deaths vary, but in South America, mortality rates from yellow fever have been as high as 80 percent. Two hundred thousand people worldwide are infected each year, but the number of actual cases is thought to be 10- to as much as 250-fold higher due to underreporting or misdiagnosis.

At one time in history, there were several factors in play that led to recurring, explosive epidemics of yellow fever. They were modes of transportation, populations of vulnerable hosts, warm weather cycles and the colonization of *Aedes aegypti* in urban environments. The circumstances today in Africa and parts of South America mirror those of Memphis in 1878 — there is poverty, people living in shanty houses, poor sanitation, containers of water in place of plumbing, a tropical atmosphere and a high population of *Aedes aegypti* mosquitoes. As the human population increases rapidly, so does the number of nonimmune people. And just like the paddleboats and trains of the nineteenth century, we now have shipping containers and airplanes. Global warming has broadened the range of disease-carrying mosquitoes. And we have a new threat: Yellow fever is listed among the pathogens that might be used during a bioterrorist attack.

* * *

Will urban outbreaks of yellow fever become more widespread and more deadly? If recent statistics are any indication, they will. Africa and South America are already moving in that direction. And unlike a virus such as smallpox, yellow fever is passed between insects and animals—it can never be eradicated because nature itself gives the virus sanctuary. Migration of humans closer to jungles and forests will place people and the virus at even closer range. But with education and routine vaccine use, the outbreaks could be contained and infect fewer people. Programs are already under way to include yellow fever in childhood vaccines in Africa.

Science is also looking for ways to understand the virus better. Recent studies have unraveled some of the mystery as to how a virus like yellow fever interacts with the human immune system. In one study, scientists identified the protein on the virus coating that interferes with the immune response. In another, scientists located part of the viral protein that human antibodies lock onto to defeat it. Isolating and reproducing that protein could lead to a safer vaccine against yellow fever.

The likelihood of the American plague returning to the United States is anyone's guess. In a 1996 article in the *Journal of the American Medical Association* the author wrote: "Because *A aegypti* mosquitoes are once again established in urban areas . . . there is widespread concern that yellow fever could erupt in explosive outbreaks, which could also occur in the southeastern United States." But we are certainly better off than people were 150 years ago. We have a vaccine. We have modern amenities like airconditioning instead of open windows, cars instead of open-air wagons. We have insect repellent. And best of all, we have the knowledge that the virus is spread by mosquitoes.

Still, viruses have taught us one thing throughout history, and it is this: That their will and ability to survive may be stronger than ours.

EPILOGUE

Elmwood

Of course there are elms at Elmwood, though they were planted after the fact to complement the name. Their massive, gnarled trunks rise high above the earth, and their roots spread deep beneath the ground, branching out amid the bones. There are also oaks. And there are magnolias with hard-shell leaves curling along the limbs, raining the dead ones like petals. It is quiet in the way that only those vast, old cemeteries can be. The only sound is the wind gathering leaves and the train that runs along tracks that edge the property.

The term *burial* brings to mind something hidden and covered, but the word *cemetery* comes from Greek and means "sleeping chamber." It's a softer approach to death, and cemeteries historically came to be places of serene recreation. Trees, flowers and streams became the natural monuments to match the stone ones. Family members once bought tickets and rode streetcars to visit the resting places of loved ones. When cemeteries moved away from city churchyards, they sprouted in the surrounding countryside, and the idea of returning to the earth what once belonged to her became all the more fitting. Over time these towns of the dead came to reveal a city's history, its stories etched in stone.

More than 125 years have passed, and still, wreaths of fresh flowers stand in Elmwood, browned and crisped by the September sun, on the tombstones of Charles Parsons, Louis Schuyler and the Martyrs of Memphis. It is a reminder—they have not been forgotten.

Old roads with names like Toof, Porter and Wellford wind through the grounds of Elmwood, where a soldier from the American Revolution and Civil War generals are buried. Tombstones, new and old, pockmark the grass like a garden of granite and marble. Some are grand, tall, ornate. Stone angels and monuments of all shapes and sizes stand in the geometry of sunlight and shade. Others are smaller, no more than two feet long, where small children have been buried.

In the middle of the cemetery is a grassy plane, strangely vacant. There are no granite tombs or crumbling concrete, just a sun-washed, treeless patch of green known as "No Man's Land." Here, 1,500 unidentified bodies are buried. At one time, their skin burned with yellow fever; now they lie in a cool, dark place where long ago their arms and legs, hands and feet, were intertwined for eternity.

Dr. William Armstrong is buried along Park Avenue in the ground beside his wife, who died on the same date as her husband, September 20, forty-six years later. Their children lay around them. A few feet away is the monument for Gideon Johnson Pillow, the general under whom Armstrong served as a surgeon during the Civil War.

Up the grassy incline from Armstrong is a flat pyramid of stone. On the four sides of the pyramid it reads *Constance, Thecla, Frances* and *Ruth*. The dates follow one another in quick succession— September 9, September 12, September 17, October 4. The point of the pyramid is the year 1878, and their bodies are buried in the shape of the cross, their tombstone standing at center.

Across the road from "No Man's Land," a tall cross atop a monument reaches heavenward. The cross has been mottled by time, streaked by years of rain. Two names and dates are carved into the stone, but the inscription that reads *priests* and *died of yellow fever* has grown shallow with age. Here, Charles Carroll Parsons and Louis Schuyler are buried together. One lived in Memphis for years surrounded by family and parishioners; the other lived in Memphis only ten days.

On November 1, 2005, the superintendent, Sunny Handback, retired from Elmwood. He had worked there since he was sixteen years old. He had scattered dirt across countless graves, occasionally meeting an old man or woman visiting the cemetery who would tell stories about yellow fever and the year 1878, when wagons full of bodies arrived, and citizens just walked into the cemetery, a corpse thrown over their shoulders and a shovel in their hands, to bury bodies anywhere they could find space. Even in recent years, groundskeepers have dug into a plot only to find the bones of an unmarked yellow fever victim buried there.

In 1878, another man held the same position as Handback. He worked as the superintendent of Elmwood during the yellow fever epidemic, and he lived on the grounds with his daughter, Grace, the "Graveyard Girl." In the cemetery's red leather logbook, the handwritten names begin in August. At first, the cause of death is listed as yellow fever, but by September of 1878, ditto marks are used, page after page. In many cases, a whole family—husband, wife and all of their children—are listed in a long row. It was Grace's hand that wrote the names, dates and cause of death, and it was Grace who rang the bell each time a body was buried. The bell tolled continuously until Grace too was stricken by yellow fever.

A burgeoning river city once stood at one of the widest points of the Mississippi River. Andrew Jackson, James Winchester and John Overton named it Memphis after the ancient, wealthy city along the Nile. Memphis, Tennessee, was a city rich in land and promise, where trains linked it to the East and West and paddleboats tied it to the North and South. It was visited by presidents and royalty, and it held the most extravagant Mardi Gras parades ever seen. White marble buildings stood on the bluff above cotton-laden steamers, and a population of white and black, northern and southern, immigrant and native saw their future. It seemed bright and certain. That city no longer exists.

The heavy German and Irish immigrant populations are gone for the most part, and the city's character has instead been shaped by the rural influence of freed slaves and farmers. Where mansions once stood along Beale Street, there is now a rough-edged, gospel-laced music known as blues. Barbecue, the food that originated in the fire pits outside slave quarters, is a culinary favorite. Many old buildings surrounding Court Square and downtown are today hollowed out with broken glass or restored as condominiums. The Gayoso Bayou now runs beneath the paved city streets.

And yet someone from 1878 would be surprised to find that many of the same contrasts remain: There is still racial strife, which reached its peak with the 1968 assassination of Martin Luther King Jr. There is still a great divide between the wealthy and poor. There are undercurrents of political corruption. There is a strong religious influence, primarily Protestant. Paddleboats still bob at the edge of the Mississippi River, and Cotton Row stands along Front Street. The Pinch District is thriving, and the Peabody Hotel is still in operation. Court Square has been restored. The city is still a major hub with boats, trains and now, planes. And the defining characteristic of the city is still a steadfast, stubborn will to survive—one that started with the devastation of the 1878 yellow fever epidemic.

Memphis was one town, one place, where yellow fever took its greatest toll, nearly destroying the city and forever changing its future, but there were hundreds more over two centuries that suffered from the American plague. Shades of those epidemics changed populations, commerce, cities, politics, wars and ultimately history. Federal laws were born in its wake. It spawned racism and prejudice, but it also inspired sacrifice and martyrdom. It created a national hero in Walter Reed and a Nobel Prize winner in Max Theiler. It touched the lives of politicians like George Washington, Thomas Jefferson, Abraham Lincoln, Rutherford B. Hayes, William McKinley and Theodore Roosevelt. It influenced literature through the likes of Washington Irving, Edgar Allan Poe, Samuel Taylor Coleridge, Sir Walter Scott, Stephen Crane and Mark Twain. And it took the lives of countless doctors, nurses, priests, nuns and ordinary civilians—most of their names have been forgotten. The American plague has been forgotten.

But in Memphis it still lives, quietly, in the bones beneath the branches of elms and in a lissome, lyre-marked mosquito that waits for the virus to find it once again.

Acknowledgments

Though I never had the honor of meeting any of the people in this story, I admire them above and beyond what could be expressed in the pages of this book. Whether the martyrs of Memphis or the martyrs of science, their courage, suffering and sacrifice are almost unmatched in today's world.

As long as I live in Memphis, I will see the ghosts of this story in the doorways of churches, on the street corners of the Pinch, along Adams and Main and in the gravestones of Elmwood. Likewise, my admiration knows no bounds for the scientists who sacrificed so much in the fight against yellow fever: Jesse Lazear, Walter Reed, James Carroll, Aristides Agramonte, Carlos Finlay, Max Theiler, among others. In an age where heroism can be so hard to come by, the fourteen human volunteers in Walter Reed's experiments amaze and inspire me.

In writing about history, a book is only as good as its research.

For their help, time and support, I thank those in the Memphis and Shelby County History Room at the Memphis Library, particularly Patrica LaPointe. Not only did she help me make some of the connections vital to this story, but she granted me access to so many irreplaceable, original documents. I also want to thank Joan Echtenkamp Klein and Claudia Sueyras with the Philip S. Hench Walter Reed Collection at the University of Virginia's Claude S. Moore Health Sciences Library. Their Walter Reed Collection is beautifully maintained; I rarely needed their assistance and that is a tribute to such an organized and accessible historical collection.

Also deserving of recognition are the curators of the Mississippi Valley Collection at the University of Memphis, the Health Sciences Library at the University of Tennessee in Memphis, the New York Academy of Medicine, the Library of Congress and the National Library of Medicine, as well as Georgia Fraser at Elmwood Cemetery, Elizabeth Wirls at the St. Mary's Episcopal Cathedral in Memphis, Professor Gary Lindquester in the Rhodes College Department of Biology and Ron Brister at the Memphis Pink Palace Museum.

One of the most valuable lessons I have learned as a writer is to be a reader of great writing. I have had the privilege to know and learn from some truly great writers: Candice Millard Uhlig, a gifted author and cherished friend; Hampton Sides, who was kind enough to give a first-time author and fellow-Memphian his help and encouragement; and Robert M. Poole, whose talent as an editor is exceeded only by his talent as a writer. A former executive editor at *National Geographic*, Bob Poole saw enough potential in me as a writer to give me a chance. I thank him for that.

I would also like to thank Mary Collins, a professor at the Johns Hopkins Zanvyl Kreiger School of Arts and Sciences, who has pushed me, challenged me and taught me. She was involved in the proposal for this book as well as in editing early drafts. I would also like to thank David Everett, my thesis advisor in the Hopkins writing program who first introduced me to the genre of narrative nonfiction.

So many friends offered their encouragement and support during this project. I would like to thank personally Allison Cates, Claire Davis, Jennifer Fox, Tessa Hambleton, Davida Kales, Lauren Kindler and Margaret McLean. Special thanks to Andy Cates, a long-suffering champion of my writing.

I am eternally grateful to my parents, Tom and Betsy Caldwell, and my in-laws Glenn and Nancy Ann Crosby, for their unceasing support, time, encouragement and commitment. I am indebted to them for allowing me to have a career in writing as well as a family without giving up one for the other. My parents have taught me to follow my passion and never once questioned where it might take me. I thank them for instilling in me such a valuable lesson—that life is too short to spend it doing anything other than what you love. As Memphians, Nancy Ann and Glenn Crosby took an active interest in this story; as a doctor, Glenn Crosby allowed me to sit in his study and riffle through medical texts; and as my in-laws, they offered their steadfast encouragement.

I would also like to thank other family members who have been sounding boards and sources of strength. My sister Lindsey has been an ever-present and perpetual believer in me, and I thank her for her unconditional love and support. Scott Crosby, an avid reader of nonfiction, always seemed sure that I would write this book. I thank him for his optimism and belief in me. Likewise, other family members have been stalwart sources of support and encouragement: Glenn Crosby, Liz Crosby, Meg Crosby and Elizabeth Crosby.

Special thanks to Mark Crosby for his photographic talents. I owe a debt of gratitude to Mark for accompanying me to Cuba in search of forgotten places. The trip would not have been the same without him; the book would not have been the same without him.

I am greatly indebted to my agent, Ellen Geiger, who was willing to take a chance on an unknown author. More than simply believing in me, she persisted until this story found its rightful place. Without her loyalty and energy, this book would never have happened. I am also grateful to my editor, Natalee Rosenstein, for her confidence in me. She took a leap of faith, and I thank her for it.

I will be forever grateful to my husband and two daughters for their own sacrifices during this project. They were patient participants in a lengthy, involved and often-chaotic process. My daughter Morgen, a master of self-expression, reminds me daily of the gift of storytelling. I thank her for giving up some of her mother to this book. Keller, who was born midway through this project and learned to be lulled to sleep by the sound of typing, has been a quieter, but no less effective, source of inspiration and encouragement. Finally, my husband, Andrew, has been unfailing in his support of me since the day we met. I am forever indebted to him for believing in me—as a writer, as a wife, as a mother. Thank you.

Notes

The introductory quote is from John Edgar Wideman's *Fever,* part of his collection first published by Henry Holt and Company in 1989.

Prologue: A House Boarded Shut

My account of the Angevine family and their deaths from yellow fever in 1878 is based on two primary sources. One is a letter written by Ray Isbell in 1978 to the *Press-Scimitar* newspaper. Lena Angevine Warner was the great-great-aunt of Isbell. Isbell recounted family stories of how an old slave investigated the house, breaking open a window, and found the corpses of the Angevine family in their various states of decomposition. The Isbell letter also described how the slave saved Lena, who was a child at the

time. Isbell's letter is part of the Eldon Roark Papers held in the
Mississippi Valley Collection at the University of Memphis.

The second source is a letter written by Lena A. Warner in
1904. She tells of her father being robbed and choked while she
was too ill to help. She also describes her experience as a nurse
during the Spanish-American War. The Warner letter is held in
the Lena Warner file of the Memphis Library, Memphis Historical Collection.

Biographical information about Lena Angevine Warner was
collected from various newspaper sources, including a 1948 obituary from the Associated Press, a 1948 obituary in the Knoxville
News Sentinel, a 1953 story in the Memphis *Commercial Appeal* and a
1994 article by Perre Magness in the *Commercial Appeal.* Biographical information is also available in Patricia LaPointe's *From Saddlebags to Science,* E. Diane Greenhill's *From Diploma to Doctorate:
100 Years of Nursing* and Paul Coppock's *Memphis Memoirs.*

There were a number of discrepancies in the facts surrounding Lena Angevine Warner, especially involving her marriage and
her role in the Walter Reed discoveries. One source wildly
claimed that Lena Warner delivered a Cuban baby, passed her own
kidney stones and performed a circumcision, a tonsillectomy and
an amputation with a kitchen knife—all in one night. In this
book, I adhered to the facts presented by Warner, her family or
those who worked with her. When a fact could not be verified by
another source, I said as much or left the material out of the book.

Part I: The American Plague

To recreate the path yellow fever followed out of Africa and across
the Atlantic, I studied the virus's behavior today. The process by
which the mosquitoes lay eggs in the hollows of trees and how the
virus was first transmitted from mosquitoes to monkeys to men

entering the West African forests for logging was based on research from two main sources: Andrew Spielman and Michael D'Antonio's book *Mosquito* and Michael Oldstone's *Viruses, Plagues, and History*. In both Africa and South America, yellow fever follows a similar course today.

Scientists generally agree that yellow fever originated in West Africa in any number of countries—where it still exists today in its purest genetic form. I chose to focus on Nigeria because that country is currently considered the hotbed of yellow fever. Descriptions of Nigeria, its plant life, topography, trade and weather, including the southwest monsoon, are based on a series of country studies published by the Federal Research Division of the Library of Congress and are available on-line or in hard copy.

I based my description of viruses in general, as well as the specifics of the yellow fever virus, on the book *Epidemic!*, which was edited by Rob De Salle and published for the American Museum of Natural History. I also relied on virus descriptions from John M. Barry's *The Great Influenza* and Gina Kolata's *Flu*. Both books do an excellent job of taking a complex subject and presenting it in comprehensible terms. For the specifics of the yellow fever virus, I studied information provided by the Centers for Disease Control and Prevention and the World Health Organization. For descriptions of the way the yellow fever virus reacts to a human cell, I relied on material from the National Institutes of Health. All of the technical information aside, personification of the virus—the idea that the virus itself is evolving, thinking, trying to conquer—is obviously a creative technique of my own making. There is no scientific evidence to suggest such.

For information about the slave trade—the Middle Passage—I based my descriptions on Madeleine Burnside and Rosemarie Robotham's book *Spirits of the Passage*. Their book not only provides general statistics about the trade but also illustrations and firsthand accounts that include some of the more disturbing details like the fact that Europeans might taste the sweat of slaves as a test for disease or that sharks trailed slave ships waiting for bodies to be thrown overboard.

The idea that yellow fever altered the history of the United States is not a new idea; after all, the virus's moniker "the American plague" says it all. Margaret Humphreys, in her book *Yellow Fever and the South*, writes, "Tuberculosis, smallpox, or typhoid might well kill as many or more every year yet fail to stir the public from apathy . . . Yellow fever was a disease whose presence often created mass panic, a response that brought commercial interactions to a standstill." To support my argument that it shaped our country's history, I compiled statistics from a wealth of sources.

Basic statistics about the number of countries and states stricken with yellow fever, as well as the number of people afflicted, were taken from the *Conclusions of the Board of Experts authorized by congress to investigate the yellow fever epidemic of 1878*. The report was written in 1879 and is available in the Rare Book Collection of the Library of Congress. The report also estimates the cost of the 1878 epidemic as $200 million, which today would be calculated as over $350 million. The reason why yellow fever has never afflicted Asia despite the right climate and the right mosquito is a mystery. Robert S. Desowitz, a professor of tropical medicine and author of *Who Gave Pinta to the Santa Maria*, suggests that it may be due to the fact that the African slave trade never extended to that part of the world.

The quote about yellow fever striking the Atlantic and Gulf states with more force than the one that bombed Pearl Harbor was taken from J. L. Cloudsley-Thompson's *Insects and History*.

The suggestion that yellow fever was the most dreaded epidemic disease for 200 years comes from Khaled Bloom's *The Mis-*

sissippi Valley's Great Yellow Fever Epidemic of 1878. Other historians have given similar opinions, and a number of doctors serving during the Memphis epidemic, and later in Cuba, offered the same impression.

That yellow fever was directly linked to the slave trade can be traced as far back as the mid-nineteenth century. In her article, "Yellow Fever: Scourge of the South," published in *Disease and Distinctiveness in the American South,* Jo Ann Carrigan writes: "Some abolitionists suggested that yellow fever was not only the result of slavery, having been introduced by the African slave trade, but that the disease served as a penalty or punishment, afflicting those areas where the institution prevailed." It was in Carrigan's article where I found the statement that yellow fever ceased in the North about the same time that slavery was abolished there. Henry Rose Carter, a friend and colleague of Walter Reed, also traced the history of yellow fever to West Africa and was one of the first to suggest that it made its way to North America through the slave trade in his 1931 book, *Yellow Fever: An Epidemiological and Historical Study of Its Place of Origin.*

The theory that yellow fever seemed divinely directed is based on some of the beliefs at the time. It was not uncommon for people to attach greater meaning to epidemics of disease—it still happens today. Even the word *plague* implies punishment in biblical terms.

According to the Greenwich Village Society for Historic Preservation, "the Village" in New York enjoyed a certain amount of seclusion until epidemics of yellow fever and cholera hit the city in 1799, 1803, 1805 and 1821. Temporary housing and businesses sprang up. The 1822 fever epidemic was an especially virulent one, and many New Yorker's settled in "the Village" for good, finally adjoining it to New York City.

The fact that Napoleon lost 23,000 troops to yellow fever in

Haiti and sold his Louisiana holdings to Thomas Jefferson, wanting to abandon conquests in this pestilent place, is taken from
Desowitz's book.

The reference to yellow fever as one of the country's first
forms of biological warfare comes from a *Washington Post* article by
Jane Singer, "The Fiend in Gray," about Dr. Louis Blackburn. I
also used information from a 2002 article in *The Canadian Journal
of Diagnosis* entitled "The Yellow Fever Plot: Germ Warfare during
the Civil War."

The impact of yellow fever on the Spanish-American War
comes from a number of sources, including the Philip S. Hench
Walter Reed Collection, held at the University of Virginia; G.J.A.
O'Toole's book *The Spanish War* and Hugh Thomas's *Cuba or The
Pursuit of Freedom,* as well as personal correspondence of the surgeon general, Theodore Roosevelt and William McKinley, among
others.

The timeline of yellow fever in North America—its prevalence in the northeast and its long reign in the South—comes
from Desowitz.

Theories about why the 1878 yellow fever epidemic proved so
deadly have appeared in a variety of publications. Many historians
have simply responded that we don't know why it was such a
deadly epidemic. In this book, I put forth the idea that it was the
combination of an El Niño cycle, an increase in new immigration
and transportation and the theory that the virus may have arrived
on ships directly from Africa rather than making its way from endemic areas in South America.

Information about yellow fever and El Niño came from an article, "A Possible Connection between the 1878 Yellow Fever Epidemic in the Southern United States and the 1877–78 El Niño
Episode," published in the *Bulletin of the American Meteorological
Society* in 1999. The article includes a timeline of El Niño cycles

during the nineteenth century; nearly all coincide with major outbreaks of yellow fever. The World Health Organization also considers El Niño weather cycles a factor in the spread of yellow fever (WHO report *Yellow Fever,* 1998).

The reference to hyacinth blooms in January comes from Bloom's book, as well as personal observation. I know that Memphians began complaining about mosquitoes based on newspaper clippings from January 1878.

The theory that Memphis was poised for greatness before the 1878 epidemic is cited in several Memphis history books. It was second only to New Orleans in population. It had survived the Civil War with very little damage. Businesses proliferated. It was the largest inland cotton market. Even the fact that Jefferson Davis chose Memphis as his home after the Civil War seems to support that idea. Unfortunately, there were also circumstances that would make Memphis vulnerable to an epidemic: poor sanitation, no clean water supply and misguided politicians.

I found the dramatic statistic about the 1878 epidemic in Memphis taking more lives than the Chicago fire, San Francisco earthquake and Johnstown flood combined in the Memphis *Avalanche* as well as in Bloom's book.

The quote that yellow fever is more calamitous to the United States of America above all other countries comes from the report of the Board of Experts, 1879, held at the Library of Congress.

Part II: Memphis, 1878

Carnival

The Edgar Allan Poe quote from "The Masque of the Red Death" is considered by some historians to be a reference to yellow fever. Poe was living during the time period when yellow fever plagued

so many cities, and the red death may have alluded to the bleeding
common from yellow fever, which is a hemorrhagic fever. The
poem tells the story of a king who locks his people away in a castle
to prevent disease. Celebrating his victory over epidemic, he
throws a lavish masque, only to find that death has indeed made
its way into the castle wearing a mask. I thought the allegory was
a chilling and perfect introduction to Memphis and its Carnival in
1878.

The description of the Mardi Gras invitation from 1878
comes from visiting the Pink Palace Museum's Memphis History
exhibit. Though they only have a few invitations from the years
that Mardi Gras took place, 1878 happens to be one of them. The
museum also displays illustrations of the Mardi Gras parades from
Harper's. Remarks about the number of people who attended the
parades, including the president of the United States, are from
the Mardi Gras file held in the Memphis History Collection of
the Memphis Library. The file contains various clippings, descriptions and newspaper illustrations. I also used an article entitled
"History of the Memphis Cotton Carnival" in the *West Tennessee
Historical Society Papers*. To recreate the 1878 parade, I read the
February and March issues from 1878 of the Memphis *Appeal*
(later to become the *Commercial Appeal*) and the Memphis
Avalanche. The majority of the details that re-create the 1878
Mardi Gras for this book came from those sources. Not only did
they give lengthy descriptions that today would seem trivial and
heavy-handed, but in reading advertisements in the newspapers, I
could piece together where shops were located on Main Street or
Second Street, what sort of clothes people wore and the fact that
caramels were sold at one store and kid gloves at another. The
newspaper is also where I found the impressive fact that the fountain in Court Square flowed with champagne during 1878 Mardi

Gras or such quaint details as the feathers escaping from ladies' fans during the ball.

For an accurate account of the weather on those two days, as well as for the remainder of 1878, I read the *U.S. Department of Agriculture's Weather Bureau from the Memphis Station Records.* I would refer to that source again and again to determine if it rained, what the clouds looked like, if a light frost fell or what the temperature might be on a given day. The weather bureau reports are held at the University of Memphis as part of the Mississippi Valley Collection.

For information about Colton Greene and the Mystic Memphi, I again went to files held in the Memphis History Collection of the Memphis Library. In those files, I found biographical information—including a description of Greene looking like Stalin—and the small details that give personality to Greene—the card allowing him admittance to the Vatican and a copy of Greene's Last Will and Testament. Likewise, I learned about the Mystic Memphi from those files, including a 1933 newspaper interview in which J. M. Semmes reminisced about the secret society that answered to the letters UEUQ.

I based my summary and description of Memphis history on three very good books written by Memphis historians: Gerald M. Capers's *The Biography of a River Town, Memphis: Its Heroic Age*; Paul Coppock's *Memphis Memoirs*; and Charles Crawford's *Yesterday's Memphis.* I also included material from Carole Ornelas-Struve and Joan Hassel's *Memphis, 1800–1900, Volume III: Years of Courage* and William Sorrels's *Memphis' Greatest Debate; a Question of Water.* The quote about Memphis refusing to take the trouble to distinguish between prosperity and progress came from Sorrels's book. A few of the specifics—like the fact that Memphis had 115 saloonkeepers—came from newspaper reports at the time.

Information about Charles C. Parsons is held in boxes as part of the Yellow Fever Collection of the Memphis Library. The boxes include letters that he wrote to his wife, as well as general opinions of Parsons. A fellow classmate from West Point described the sort of fanaticism evident on Parsons's face. Men who served with him during the Civil War described his courage at the Battle of Perryville. There is even a letter from Jefferson Davis to Parsons. I found the sermon Parsons gave on the eve of the 1878 Mardi Gras in a scrapbook that belonged to George C. Harris, whose papers are also held in the Yellow Fever Collection at the Memphis Library. The sermon had been printed in the *Ledger* newspaper in February 1878, and Harris kept it for his scrapbook. Although I only included part of it, the full sermon can be found in the Harris papers.

Very few photos from that decade exist. In order to create a visual sense of downtown Memphis from a visitor's point of view, I studied an 1870 map of Memphis drawn by the U.S. Departments of Agriculture and Commerce. The original map is held at the Library of Congress, though copies are in wide circulation. The Mississippi Valley Collection at the University of Memphis has vignettes of the Victorian houses along Adams Street—an area now known as Victorian Village—in the Eldon Roark Papers. I also studied architectural drawings of Memphis buildings. I was surprised to find that in addition to the wooden and brick buildings one would expect, Memphis had several grand structures designed by prominent architects. The columns of the Gayoso Hotel, the building-in-progress of the Customs House, the glass-covered Water Works, and an elaborate prison, among others, would offer a stunning view from the river. None of those buildings exist in Memphis today—only the gates of the prison are still standing near the entrance of Mud Island.

I relied on descriptions of the Greenlaw Opera House and the

Exposition Building from articles in the West Tennessee Historical Society Papers, an excellent resource for details and specifics
about any number of subjects relating to Memphis history.

Other sources consulted for descriptions of Memphis, photographs or illustrations were Robert A. Sigafoos's *Cotton Row to
Beale Street*, Beverly G. Bond and Janann Sherman's *Memphis in
Black and White*, Robert W. Dye's *Images of America: Shelby County*
and Ginny Parfitt's *Memories of Memphis: A History of Postcards*.

Bright Canary Yellow

It was widely believed in 1878 that the yellow fever epidemic could
be traced to the steamer *Emily B. Souder*, which sparked a number
of cases in New Orleans. For a physical description of the *Emily B.
Souder*, its history and to learn its fate, I looked up *American Lloyd's
Register of American and Foreign Shipping* (1865) and the *Record of
American and Foreign Shipping* (1871)—originals of both documents
have been scanned and are available on-line. During that time period, news of a ship's landing and departure was also printed in the
newspaper. As the *Souder* sailed out of New York, I found references to the ship in the *New York Times* and ultimately found the
most fateful one: the *Souder* sank in December 1878.

For my account of the *Souder*'s trip to New Orleans, the
deaths of John Clark and Thomas Elliott and the autopsies, I re-
lied on two primary sources: "History of the Importation of Yel-
low Fever into the United States, 1693–1878," presented by Dr.
Samuel Choppin at the meeting of the American Public Health
Association on November 21, 1878, and the "Report upon Yellow
Fever in Louisiana in 1878," by Dr. S. M. Bemiss in the *New Orleans
Medical and Surgical Journal* (1883). Information about the *John D.
Porter* was also taken from Bemiss's report and J. M. Keating's ac-
count of the epidemic.

I found corroborating information from additional sources: Khaled Bloom's *The Mississippi Valley's Great Yellow Fever Epidemic of 1878*, J. H. Ellis's *Yellow Fever and Public Health in the New South* and Jo Ann Carrigan's *The Saffron Scourge*. It was in Choppin's own report to the American Public Health Association that I found his statement about Thomas Elliot's death: "These are all the usual appearances observed in the examination of a person dead of yellow fever, and we had no doubt that the man had been the subject of this disease." Prior to that, Choppin had claimed that he had no idea the crewmembers of the *Souder* had yellow fever; he also denied that the subsequent yellow fever outbreak had anything to do with the *Souder*'s May arrival. Using Choppin's own paper and details from the minutes of the Memphis Board of Health, held in the Memphis History Collection of the Memphis Library, I pieced together the timeline in which Choppin and New Orleans officials were first aware of yellow fever cases and when Memphis was officially notified two months later.

My account of the *Aedes aegypti* mosquito and its behavior was based on Spielman and D'Antonio's *Mosquito*, Jerome Goddard's *Physician's Guide to Arthropods of Medical Importance* and Carlos Finlay's studies of the mosquito.

Information about the prevalence of yellow fever in Cuba came from Henry Rose Carter's book, and statistics about the marked virulence of the 1878 epidemic can be found in Jo Ann Carrigan's book, as well as Humphreys's and Bloom's.

The Doctors

To construct what downtown Memphis would have felt like in 1878, I relied on a book written by the Reverend D. A. Quinn, *Heroes and Heroines of Memphis or Reminiscences of the Yellow Fever Epidemics*. The book, published in 1883, is part of the Yellow Fever

Collection at the Memphis Library. In it, I found detailed descriptions of Court Square, the flowers blooming there, women pushing baby carriages, bootblacks, the milkman's morning cry "Wide Awake!" and the newsboys shouting headlines.

For descriptions of the weather—namely the drought and heat—I used newspaper clippings from 1878 and the *U.S. Department of Agriculture Weather Bureau, Memphis Station Records* in the Mississippi Valley Collection at the University of Memphis. Details about the raw sewage and dead animals came from J. M. Keating's *A History of Yellow Fever: The Yellow Fever Epidemic of 1878.* Keating was the editor of a local newspaper and survived the epidemic. A year later, he published the definitive book on the subject.

Information about medicine in the nineteenth century came from a variety of sources: W. F. Bynum's *Science and the Practice of Medicine in the Nineteenth Century,* Thomas J. Schlereth's *Victorian America* and Paul Starr's *The Social Transformation of American Medicine.* For more specific information about medical practices in Memphis, I relied on Patricia LaPointe's book, *From Saddlebags to Science.* I found references to medications like Tutt's pills or doctors specializing in "secret diseases" in 1878 newspaper clippings.

For further explanation on the contagionists versus noncontagionists and the local versus exotic origin of yellow fever, see Simon R. Bruesch's article in the *Journal of the Tennessee Medical Association,* Margaret Humphreys's *Yellow Fever and the South* and Margaret Warner's "Hunting the Yellow Fever Germ" in the *Bulletin of Historical Medicine* (1985). Details about the history of quarantines was taken from Keating's book. Details about the Quarantine Act came from John Ellis's *Yellow Fever and Public Health in the New South.*

Information about "the war of the doctors" and the activities of the Memphis Board of Health during June, July and August, was found in the minutes recorded at their meetings, which are held in the Memphis History Collection of the Memphis Library. I also followed "the war of the doctors" in the Memphis *Appeal* and in a 1978–79 article in the *Journal of the Tennessee Medical Association* by Dr. Simon R. Bruesch, whose collection of materials is held at the Health Sciences Historical Collections of the University of Tennessee, Memphis.

For a description of Dr. Robert Wood Mitchell, I looked to the Simon Rulin Bruesch Collection. I based my descriptions of John Erskine on the Erskine file in the Memphis History Room of the Memphis Library, as well as Bruesch's article.

The quote about privies and the general state of water in Memphis during that time period was taken from Sorrels's book, held in the Mississippi Valley Collection at the University of Memphis.

"Memphis is about the healthiest city on the continent at present" was printed in the Memphis *Appeal*, June 22, 1878.

"Is it not better to expend a few thousand as a safeguard than lose millions . . . besides the thousands of valued lives that will have passed away" appeared in the Memphis *Appeal*, July 4, 1878.

The amount of money—$8,000—secured at the July 6 meeting of the Board of Health to clean up the city was reported in the Memphis *Appeal*, July 6, 1878.

"Should an epidemic reach Memphis . . . those who opposed the establishment of a quarantine will be held responsible" was printed in the Memphis *Appeal*, July 11, 1878.

Mitchell's letter of resignation from the Memphis Board of Health appeared in the *Appeal*, July 11, 1878.

"The yellow fever scare is about over in Memphis" was printed in the Memphis *Appeal*, July 30, 1878.

The details of Dr. John Erskine boarding the *John D. Porter* for inspection appeared in the Memphis *Appeal*, July 30, 1878.

Information about the strange occurrences in July of 1878—the streetlights exploding, the Edison speaking phonograph, the rattlesnake, the cocktails and the eclipse—were all taken from the Memphis *Appeal* and *Avalanche* newspapers.

The description of the constellation Ophiuchus was based on information in *Encyclopaedia Britannica*. Imhotep, the man on whom the constellation is based, was called by Sir William Osler "the first figure of a physician to stand out clearly from the mists of antiquity." It is said that the medical sign of two serpents coiled around a staff is based on Imhotep.

The timeline of the first yellow fever cases and the description of the mass exodus out of Memphis were taken from Keating's book. Additional information about William Warren was found in 1878 copies of the newspapers. Descriptions of the Pinch District were taken from files by the same name held in the Mississippi Valley Collection of the University of Memphis and the Memphis Historical Collection at the Library of Memphis. And specifics about Bionda's snack house in the Pinch were found in Thomas Baker's article "The Yellow Fever Epidemic of 1878 in Memphis, Tennessee" in the *Bulletin of the History of Medicine*, 1968.

The *New York Times* editorial about New York's filthy tenement houses appeared in John Ellis's *Yellow Fever and Public Health in the New South*.

The quote about the tales circulated by sensationalists was from the *Appeal*, August 4, 1878.

The estimated number of people left in Memphis during the epidemic and the number stricken is based on Keating's book. According to Keating, 19,600 remained, and 17,600 had yellow fever. According to Baker's article, 25,000 people had fled the city—over half of the population—in the four days after Bionda's death. "For the sake of humanity, men became inhuman" was taken from Keating's book.

The story of the Memphians released from the trains in Milan
for provisions appeared in an *Evening Appeal* clipping found in the
Yellow Fever Collection at the Memphis Library.

The description of the Citizen's Relief Committee and their
actions was based on Bloom's book, reports in the local newspapers and the Charles G. Fisher Papers at the University of Memphis. Information about the Howard Association was taken from
Bruesch's article.

The Board of Health's declaration of a yellow fever epidemic
on August 23 was taken from the board's minutes.

A City of Corpses

Descriptions of the city during the epidemic came from a number
of sources: Keating's firsthand account of the epidemic in his
book; Reverend D. A. Quinn's book *Heroes and Heroines of Memphis
or Reminiscences of the Yellow Fever Epidemics;* and Dr. J. P. Dromgoogle's *Yellow Fever Heroes, Honors, and Horrors of 1878.* All three
books are part of the Yellow Fever Collection at the Memphis Library. I also consulted the George C. Harris papers at the Memphis Library, Charles G. Fisher papers at the University of
Memphis and the accounts of the nuns at St. Mary's. A number of
the descriptions came from the accounts of the epidemic in the
Appeal and the *Avalanche.*

I found the quote from Rutherford B. Hayes's personal letter
calling the Memphis epidemic "greatly exaggerated" in the Rutherford B. Hayes Presidential Center on-line collection of the president's letters and diaries. The dates when the mayor of Memphis
and other officials wired the president were found in local papers.

. Nearly all information pertaining to Sister Constance and the
nuns at St. Mary's was taken from a series of notes and letters
found among Constance's personal items after her death—they

were collected and printed, though not published, as *The Sisters of St. Mary at Memphis: With the Acts and Sufferings of the Priests and Others Who Were There with Them during the Yellow Fever Season of 1878.* St. Mary's Cathedral in Memphis houses that book.

Descriptions of St. Mary's Episcopal Cathedral and information about it were based on visits there and an interview with Elizabeth Wirls. I also read the bound, printed history of St. Mary's, available at the church. While St. Mary's still has the original altar belonging to the nuns, the stole worn by Charles Parsons and the stained-glass rose window—that is almost all that remains of the original cathedral, which burned down several years after the epidemic. During a stay in Kansas City, however, I visited the St. Mary's Episcopal Church, which was built in the likeness of the Memphis church. It even has stained-glass windows commemorating the Martyrs of Memphis. It was there that I was able to get a sense of the dark-wood interior and Victorian, gothic architecture of the Memphis cathedral as it was when the nuns served there.

Some details, like the yellow cardboard hanging from doors, "bring out your dead" and the burning of infected clothing appeared in a 1932 *Press-Scimitar* clipping held in the Yellow Fever Collection at the Memphis Library.

I based my descriptions of Victorian mourning on the books *Victorian America: Transformations in Everyday Life* and *Elmwood: In the Shadows of the Elms.*

"A stranger in Memphis might have believed he was in hell" appeared in an article, "City Still Bears Scars of Epidemic Century Ago," in the *Commercial Appeal,* June 18, 1978. The article was found in the Yellow Fever Collection. Also from that article came the statistic about one railroad ticket agent who sold $35,000 worth of train tickets in three days.

Descriptions of Elmwood Cemetery during the epidemic

came from *Elmwood, History of the Cemetery,* written in 1874, and available at the Elmwood offices. I also visited the cemetery to read through its ledger book for 1878.

I found a reference to letters with holes punched through and fumigated with a studded paddle in a 1938 *Commercial Appeal* clipping in the Yellow Fever Collection of the Memphis Library.

The Destroying Angel

All information about Armstrong came from his original letters held in the William Armstrong papers in the Yellow Fever Collection and from an article in the *West Tennessee Historical Papers,* 1950. Additional information about him was found in the papers belonging to Constance and the nuns at St. Mary's, as well as the William Armstrong file held at Elmwood Cemetery.

To recreate scenes involving Constance, Parsons and Armstrong, I relied on their personal letters, already cited, and accounts of the nuns. Though Constance and Parsons often met, as did Constance and Armstrong, I could only find only one case where all three were together—the one included in this book—when Dean Harris fell ill.

Information about doctors during the epidemic as well as general information about Victorian medicine came from a variety of sources: the Breusch Papers, LaPointe's *From Saddlebags to Science,* Dromgoogle's *Yellow Fever Heroes, Honors, and Horrors of 1878,* Keating's *The Yellow Fever Epidemic of 1878,* Paul Starr's *The Social Transformation of American Medicine,* T. O. Summers's *Yellow Fever* and Goodman and Gillman's *The Pharmacological Basis of Therapeutics.* I also based some of the details like spring-loaded lancets or leaden glass bottles on visits to the Memphis history exhibit at the Pink Palace.

Descriptions of the day-to-day work of the Howard physicians were taken from Breusch's article in the *Journal of the Tennessee Medical Association*. Details about Mitchell's specific treatment for fever cases was found in Dromgoogle's account. The bizarre treatment for yellow fever involving sitting on a chair and passing out was taken from Dr. S. S. Fitch's *The Family Physician,* 1876.

The quote from the 1878 *Boston Medical and Surgical Journal* about surgeons serving in war versus the yellow fever epidemic was found in the Breusch papers.

"Only one change was noticeable—the decrease of their numbers" is from Keating's *The History of Yellow Fever.*

All information pertaining to Charles C. Parsons was taken from the Parsons file, George C. Harris papers and papers belonging to the Sisterhood at St. Mary's—all are held in the Yellow Fever Collection at the Memphis Library.

"A man on Poplar . . . cowardly deserted his wife and little daughter, both of whom were ill with the fever; if he isn't dead, somebody ought to kill him" appeared in the September 6, 1878, *Appeal.*

The account of Louis Schuyler was taken from the papers belonging to the Sisterhood of St. Mary's, as well as the George C. Harris papers. A fictional description of his death is also available in Charles Turner's *The Celebrant.* It was never confirmed, nor denied, that Louis Schuyler was moved into the death alley still alive. After his death, letters from concerned friends and family were sent to Dean George C. Harris. Those letters, as well as his explanation, are part of the Yellow Fever Collection.

The description of the illness and deaths of Constance and Thecla were taken from the papers belonging to the Sisterhood of St. Mary's. A few of the details were also found in the William J. Armstrong papers. The obituary quote, "Of them it may be said they were lovely in their lives, and in their death they were not divided," was found in the scrapbook of George C. Harris.

The account of Dr. William J. Armstrong's final days and death were taken from his personal correspondence and the article about his life and letters in the *West Tennessee Historical Papers*, 1950.

The final list of those who perished in the epidemic was taken from several sources, including Keating's book, Breusch's papers, the *Elmwood* book, the Charles G. Fisher papers, and *Memphis 1800–1900, Volume III: Years of Courage*.

Descriptions of the citizen's meeting at the Greenlaw on Thanksgiving Day was taken from newspaper accounts.

I found that the *Emily B. Souder* sank in December of 1878 in the shipping news of the *New York Times*.

Greatly Exaggerated

Statistics from the 1878 epidemic came from three sources: Bloom's *The Mississippi Valley's Great Yellow Fever Epidemic*, Carrigan's *The Saffron Scourge* and the *Conclusions of the Board of Experts, 1879*.

Information about the state of the country following the epidemic and the battle over the National Board of Health came from John Ellis's *Yellow Fever and Public Health in the New South* and Margaret Humphreys's *Yellow Fever and the South*. The quote about the president not wanting to commit the recessed congress to an investigation was taken from Ellis's book, as was the quote from *Harper's*.

Statistics about financial aid and provisions sent to Memphis appeared in Keating's book.

My account of the Board of Experts was taken from the *Proceedings of the Board of Experts* and the *Conclusions of the Board of Experts, 1878*. Both are held in the Rare Books Collection of the Library of Congress. Statistics for the number of blacks and whites who died were taken from those reports, as well as Keating's book.

The peculiar incidence of the fever among white children in New Orleans was taken from Carrigan's book.

The Havana Commission

Biographical information about Juan Carlos Finlay was taken primarily from an article, "Carlos Finlay's Life and the Death of Yellow Jack," published in the *Bulletin of the Pan American Health Organization* (1989).

Biographical information about George M. Sternberg came from the George Miller Sternberg papers at the National Library of Medicine, John Pierce and Jim Writer's *Yellow Jack*, Martha Sternberg's *George Miller Sternberg: A Biography* and "The Trials and Tribulations of George Miller Sternberg—America's First Bacteriologist," published in *Perspectives in Biology and Medicine*.

Reparations

Historians agree that 5,150 people died during the Memphis 1878 epidemic. That number comes from Keating's book written only one year after the epidemic. The number represents roughly one-tenth of the total population of Memphis, which was estimated to be just under 50,000 in 1878.

The evolution of Memphis from a cosmopolitan, diverse, progressive city to one in which Protestant fundamentalism and white supremacy flourished is taken from Memphis historian Gerald M. Capers in his book *The Biography of a River Town*. The quote suggesting that Atlanta owes its present position to the work of the *Aedes aegypti* in Memphis is from the same source.

The statistic taken from the National Bureau of Education census
is also from Capers book: "The extent to which newcomers took
the places of former residents in the years following 1880 is revealed in a census taken in 1918 by the National Bureau of Education. Of the 11,781 white parents residing in Memphis forty years
after the great epidemic, only 183, less than 2 per cent, had been
born there."

Information about George Waring and the account of his
death came from William W. Sorrels's *Memphis' Greatest Debate; a
Question of Water,* as well as his *New York Times* obituary on October
30, 1898, the *Commercial Appeal* obituary, the Memphis *Avalanche*
obituary and Waring's own report, *The Memphis Sewerage System.*

Part III: Cuba, 1900

A Splendid Little War

The introductory quote for Part III is part of a letter written by
Walter Reed to his wife, Emilie, on December 31, 1900.

The description of the sinking of the USS *Maine* was taken
from Captain Sigsbee's own book, *The Maine,* written in 1899, and
held in the Rare Book Collection of the Library of Congress. I
also relied on G.J.A. O'Toole's *The Spanish War: An American Epic
1898,* an excellent book on the war. There is information from
Hugh Thomas's *Cuba or the Pursuit of Freedom* and Pierce and
Writer's *Yellow Jack* as well.

The four presidents who attempted to purchase Cuba at one
time or another were James Polk, Franklin Pierce, James
Buchanan, and William McKinley. The quote from Thomas Jefferson was taken from *Yellow Jack.* Robert Desowitz's quote appeared in his book *Who Gave Pinta to the Santa Maria?*

Siboney

I based the majority of the "Siboney" chapter on Victor Clarence Vaughn's autobiography, *A Doctor's Memories.*

Additional information, including the Round Robin letter controversy, was taken from O'Toole's *The Spanish War.* Shafter's quote was also taken from that source. It has been disputed whether or not the Round Robin letter was Shafter's idea or Roosevelt's. *The Spanish War* uses Roosevelt's own autobiography as the source for the account in which Shafter came up with the plan and wrangled Roosevelt into it; but, in a footnote, O'Toole adds that Roosevelt had written a letter to a friend four days before the meeting with Shafter in which he enclosed a draft of the Round Robin letter.

Biographical information about George M. Sternberg came from the George Miller Sternberg papers at the National Library of Medicine, The Philip S. Hench Walter Reed Collection, John Pierce and Jim Writer's *Yellow Jack,* Martha Sternberg's *George Miller Sternberg: A Biography* and "The Trials and Tribulations of George Miller Sternberg—America's First Bacteriologist," published in *Perspectives in Biology and Medicine.* The editorial from the *New York Times* calling Sternberg unfit for the position of surgeon general appeared in "The Trials and Tribulations of George Miller Sternberg—America's First Bacteriologist."

An Unlikely Hero

The majority of Reed's biographical information came from Dr. William Bean's excellent biography *Walter Reed,* published in 1982. It is a well written and thorough account of Reed's work in Cuba. Additional information was taken from Howard A. Kelly's *Walter*

Reed and Yellow Fever, 1906; Laura Wood Roper's *Walter Reed: Doctor in Uniform,* 1943; and Pierce and Writer's *Yellow Jack,* 2005. Several personal details and excerpts from letters were taken from the Philip S. Hench Walter Reed Collection, which houses a wealth of personal correspondence between Reed and Emilie, as well as other family members, and from the Walter Reed Papers at the National Library of Medicine.

Information about Dr. Luke P. Blackburn was taken from Jane Singer's "The Fiend in Gray," published in the *Washington Post.* Some information regarding Blackburn was also found in Quinn's 1887 book *Heroes and Heroines of Memphis or Reminiscences of the Yellow Fever Epidemics.*

Historical information about Johns Hopkins University was taken from three sources: John M. Barry's *The Great Influenza,* Roper's *Walter Reed: Doctor in Uniform* and Kelly's *Walter Reed and Yellow Fever.* Kelly, one of the Great Four, worked with Reed at Hopkins.

For the account of the Typhoid Board's work, I relied on Victor Vaughan's *A Doctor's Memories.* Information about Vaughan's medical advancements after the Typhoid Board and his quote at the end of the chapter were taken from Barry's *The Great Influenza.*

A Meeting of Minds

Many of the descriptive details about the sights, smells, sounds and feel of Havana came from personal experience when I visited the city in December 2005 while researching this book.

According to historian Philip S. Hench, the military camp six miles outside of Havana was known as Camp Columbia during the Spanish-American War. Following the war, the regimental tents were replaced with wooden buildings, and the "army of occupation" moved in, renaming the camp the Columbia Barracks. I use the terms interchangeably since many of the principle characters were still in the habit of calling the barracks Camp Columbia.

To recreate scenes at Camp Columbia, I relied on several different sources: Albert Truby's *Memoir of Walter Reed: The Yellow Fever Episode*, Bean's *Walter Reed*, Pierce and Writer's *Yellow Jack*, as well as personal letters from soldiers who stayed there and Philip S. Hench's own documentation.

For details about entertainment for the enlisted men and trips to Havana, I used an interview between Philip S. Hench and Paul Tate conducted in 1954.

For the description of Walter Reed's voyage to Havana in March 1900, I relied on Philip S. Hench's interview with Lawrence Reed and Blossom Reed in 1946. Information about electrozone and Reed's work with it was taken from Reed's own report to the surgeon general, dated April 20, 1900, and Truby's *Memoir of Walter Reed*.

For information about Jesse W. Lazear, I used a number of sources, including Philip S. Hench's biographical notes on Lazear, photographs from Lazear's own photo album in Cuba, photographs from the Philip S. Hench Walter Reed Collection at the University of Virginia, letters from Lazear to his mother, a *Baltimore Sun* article dated September 9, 1905, and Aristides Agramonte's *The Inside History of a Great Medical Discovery*.

Lazear's quote that Walter Reed was another convert of the mosquito theory was found in Hench's interview with Reed's children, as well as Hench's questionnaire for Jefferson Kean in 1946.

The account of the officers in their dress whites discussing medicine on the veranda in the evenings was based on Truby's book. Truby noted that Reed's interest in yellow fever was tireless.

The details about Lazear's courtship with Mabel, his trips to California to meet her family and visit their ranch, as well as the date and place of their marriage, came from a series of letters written by Jesse Lazear to his mother, Charlotte Sweitzer, in July, August and September of 1896. Those letters are part of Hench's collection at the University of Virginia.

Lazear's description of Havana was taken from a letter to his mother on February 11, 1900. Lazear's personal photo album of their first few months in Havana and Marianao is held in the Hench collection. The letter Lazear wrote to his mother requesting that she store the boxes of golf clubs, Shakespeare, linens and dishes was dated February 15, 1900. Additional descriptions about the Lazears' home in Cuba—like a partition made of matting from a Chinese store, the shower bath and physical details about the home—were found in letters Lazear wrote to his mother during February and March of 1900. Additional details about what they ate and what they fed the baby, as well as descriptions of taking Houston to the beach were found in letters to Lazear's mother, dated February 15 and March 15, 1900.

The account of tree frogs settling into the rafters of buildings at Columbia Barracks was found in Truby's *Memoir of Walter Reed*.

On April 6, 1900, Lazear wrote to his mother to tell her that Mabel and Houston would be returning to the States on the *Sedgwick* around April 14. As transports were often a day early or a day late, that date may not be exact. Walter Reed, in a letter to his wife, makes reference to a ticket on a transport like the *Sedgwick* costing $12.

Additional details about the soldiers' time at the Columbia Barracks were gathered from various letters and photographs of the social hall. Jesse Lazear wrote to his mother that he often stood on his porch, but rarely ventured to the weekly dances. I know he could hear the firing of the hour at El Morro Castillo because Walter Reed wrote to his wife, on December 31, 1900, that they could hear it on quiet nights.

Information about Major Jefferson Randolph Kean and the diary of yellow fever cases that he kept was taken primarily from Bean's *Walter Reed*. Kean's description of Reed, his whimsical humor and penchant for quaint stories, was found in a letter Kean wrote to Philip S. Hench on January 23, 1944.

I found all information about Lazear's investigation of Sergeant Sherwood's case of yellow fever—the tests conducted and the autopsy notes—in a notebook kept at the New York Academy of Medicine. The notebook has a fascinating history. It mysteriously disappeared after Reed's work in Cuba was completed. It was discovered thirty years later in an ash barrel and sold for twenty-five dollars to Archibald Malloch, who gave it to the academy. The New York Academy of Medicine has the notebook labeled as the "Record of the Yellow Fever Commission's work in the handwriting of Dr. Neate." Neate was Walter Reed's lab assistant. However, historians Philip S. Hench and Reed biographer Laura Wood Roper located the notebook at the academy in the 1940s, and both believed the handwriting in the notebook to be that of Lazear's and Reed's. Likewise, when I visited the New York Academy of Medicine in 2004, I took samples of Lazear's and Reed's distinctive handwriting; they matched the records in the notebook perfectly. In my book, I have therefore described this notebook as the one belonging to Jesse Lazear until September 1900. In December of that year, the records began again in Walter Reed's handwriting.

The quote about entomology, keying in anatomical minutiae, the formation of mouthparts, the bewildering pattern of wing venation is from Robert Desowitz's book.

Truby's quote about the epidemic in Quemados was taken from his book *Memoir of Walter Reed*.

The account of Kean's illness and his visit to Major Edmunds is based on a letter from Jefferson Randolph Kean to Philip S. Hench on January 23, 1941.

The Yellow Fever Commission

The letter from Surgeon General George Miller Sternberg to the adjutant general on May 23, 1900, is held in the Philip S. Hench Walter Reed Collection at the University of Virginia. In the original, typed version the reference to yellow fever has been stricken by hand. The reason for this is not known, but Hench believes it may have been to avoid offending the Cubans. At that time, yellow fever was still considered a disease associated with filth.

Again, for an impression of the Havana harbor I relied on personal experience as well as descriptions from letters. The account of Reed's trip to Havana on board the *Sedgwick* was from a letter he wrote to Emilie on June 25, 1900. It is held in the Hench collection.

Information about Sanarelli and the bacteria he believed caused yellow fever was taken from Kelly's book. As Kelly was a contemporary of both Sternberg and Reed, he was well aware of the controversy.

I based the theory that Reed, rather than Sternberg, chose the members of the Yellow Fever Board on the electrozone report he filed with the surgeon general on April 20, 1900. It can be found in the archives of the National Library of Medicine. In the report, Reed specifically thanks Carroll, Agramonte and Lazear for all of their help.

Biographical information about James Carroll came from several sources. I found a reference to his nickname as "Sunny Jim" in an interview Hench conducted with Kean on November 19, 1946. Carroll himself provided a lot of the information about his background to Caroline Latimer in a letter in 1905. Latimer later wrote an article about Carroll published in *A Cyclopedia of American Medical Biography,* 1920. Additional information was taken from Bean's book.

Lazear's description of Carroll was in a letter written to Mabel on July 15, 1900. And other impressions of Carroll were taken from an interview with Pinto and the memory of Charles S. White, a former student of Carroll's and Reed's. All can be found in Hench's collection.

The quote by a colleague about Carroll needing to be led by a man with vision was in a letter Albert Truby wrote to Hench on September 3, 1941.

The letters in which Carroll chastises Jennie were written on August 27, 1900, and December 1, 1900. Both are held in the Hench collection and there are references to them in Bean's book. Not all of his letters were so negative, but those two were chosen because they reflect what seemed to be a strained relationship between Carroll and his wife, as well as his children. In many other letters, he is perfectly cordial though never very affectionate. The information about Carroll's son was found in a memorandum written by Hench on November 11, 1954.

Biographical information about Aristides Agramonte was taken primarily from the curriculum vitae of Aristides Agramonte, held in the Hench collection. There is also a photo in the collection on which I based the physical description of Agramonte. The reference to Agramonte's work with Reed in Washington prior to their appointment in Cuba was found in Hench's "Timeline of Agramonte's Service in the Army Medical Corps." Agramonte was assigned to work with Reed at the lab in Washington for the month of May, 1898.

William Welch's recommendation of Lazear was part of a letter written to the surgeon general on January 12, 1900.

Reed's account of the journey on board the *Sedgwick* was taken from his letter to Emilie on June 25, 1900. Details about the harbor, Plaza de Armas, and the Governor's Palace came from personal observation during the trip to Havana. Descriptions of the plant life came from both personal observation and Reed's descriptions in letters to his wife.

The board's meetings on the veranda and the instructions from Sternberg come from Truby's account as well as Sternberg's biography. Agramonte described the scene in full, including the reverence with which they listened to Reed. The description of Reed as a teacher comes from Hench's Notes on Reed and Carroll by Charles S. White, January 10, 1942. And, Lazear's excitement about working with Reed was found in a letter to his mother, Charlotte Sweitzer, on May 29, 1900. The specific orders from Sternberg to the board can be found in the letter from George Miller Sternberg to Walter Reed, May 29, 1900, in the Hench collection.

Sternberg's quote about mosquitoes being a useless investigation came from a letter in the Hench collection from Henry Hurd to Caroline Latimer, February 11, 1905. In the letter, Hurd relates a conversation he and Reed had in which Reed described the scene and quoted the surgeon general. The veracity of the story cannot be proven, but it does seem plausible. In what was a sad end to a skillful, twenty-year study of yellow fever, Sternberg's judgment had become clouded by the controversy with Sanarelli. In the end, Sternberg would take credit for suggesting the mosquito possibility to Reed.

Insects

The majority of the description of the Board's lab came from Albert Truby's *Memoir of Walter Reed,* 1943. A few details were taken from other sources: The fact that the lab used to be the operating room came from John Moran's account, and the description of the jars of black vomit was found in Philip S. Hench's interview with Gustav Lambert in 1946.

The finer points of Reed's first weeks in Cuba were taken directly from his letters to Emilie dated July 2, 7, 19, 20, 23, 27 and 30. Reed's reference to returning to the United States to finish the typhoid report came from a letter written to Surgeon General Sternberg on July 24, 1900. Some details were also provided by Philip S. Hench's interview with Blossom Reed in 1946. All of the letters are held in the Philip S. Hench Walter Reed Collection at the University of Virginia.

Other details in the chapter—like the pastimes for the soldiers—came from Truby's account.

The majority of the material about Lazear's time in Cuba during the summer of 1900 came from letters to his wife, Mabel, and his mother, Charlotte Sweitzer. The letters are held in the Hench collection. Additional information about his mother being twice widowed and losing two of her sons came from J. A. del Regato's "Jesse William Lazear: The Successful Experimental Transmission of Yellow Fever by the Mosquito," published in *Medical Heritage* in 1986.

Details about the visit from Dr. Herbert Durham and Dr. Walter Myers of the Liverpool School of Tropical Medicine came from a letter written by Reed to his wife on July 19, 1900, and from an article by Durham and Myers, "Yellow Fever Expedition," published in the *British Medical Journal* on September 8, 1900. Additional information about both Durham and Myers contracting yellow fever came from B. G. Maegraith's "History of the Liverpool School of Tropical Medicine," in *Medical History*, October 1972 and William Petri's "100 years of Tropical Medicine and Hygiene," in the *American Journal of Tropical Medicine and Hygiene*, 2004.

The account of Aristides Agramonte's visit to Pinar del Rio

came primarily from his own account, *The Inside History of a Great Medical Discovery*. Additional details were provided by Reed's letter to Surgeon General Sternberg on July 24, 1900.

The descriptions of Pinar del Rio were taken from Christopher Baker's *Cuba*. Although I visited the city of Havana, I was not able to venture into the countryside around Pinar del Rio.

The remarks about Aristides Agramonte not being present at the meeting when the board decided to self-experiment were published by John C. Hemmeter in the *American Public Health Reports* in 1908 under the title "Major James Carroll of the United States Army, Yellow Fever Commission, and the Discovery of the Transmission of Yellow Fever by the Bite of the Mosquito 'Stegomyia Fasciata.'" Most historians agree that the article was very one-sided in favor of Carroll. It was written by a former classmate of his. While there is a great deal of helpful information, his criticism of Agramonte seems unjust. Likewise, his reference to Reed abruptly leaving Cuba the morning after the meeting, found in a letter from James Carroll to Howard A. Kelly on June 23, 1906, is unsubstantiated. In fact, as outlined in this book, there were a number of earlier references made by Reed that indicate he planned to leave on that date. Carroll's attack on Agramonte and Reed offers further evidence of his mental state after his work in Cuba. He obviously felt that he was not given due recognition.

Reed's reference to human experimentation came from his July 24th letter to Sternberg.

The story of Reed's return trip to the United States was taken from Truby's account—including Reed's joke about the "Rollins."

Vivisection

The best book I've found on vivisection and the source for much of this chapter is Susan Lederer's *Subjected to Science*. I also consulted Lawrence Altman's *Who Goes First? The Story of Self-Experimentation in Medicine.*

Information about Edward Jenner's experiments on his son was taken from Greer Williams's *Virus Hunters.*

The study that George Sternberg and Walter Reed conducted on children in orphanages was published in 1895 in *Transactions of the Association of American Physicians* as "Report on Immunity against Vaccination Conferred upon the Monkey by Use of the Serum of the Vaccinated Calf and Monkey."

The Tennyson quote comes from his poem "In the Children's Hospital," published in *The Poems of Alfred Lord Tennyson.* The reference to the poem was found in Lederer's book.

Did the Mosquito *Do It?*

The dates surrounding the time the board first visited Carlos Finlay were kept purposely vague. Some accounts claim that the Yellow Fever Board first visited Finlay in early July, just after their arrival. Other accounts say it didn't happen until early August. There is no definitive proof either way.

Much of the details surrounding Carroll's infection came from Agramonte's account, as well as Philip S. Hench's speech "The Conquest of Yellow Fever," written on January 1, 1955. I also found details about the illness that Lazear wrote in his logbook, now held at the New York Academy of Medicine.

Lazear's mention of trying to find the real yellow fever germ rather than bothering with Sanarelli came from a letter written to wife, Mabel, on August 23, 1900. His reference to the distance seeming very great at a time like this was found in a letter to his mother dated August 27, 1900. Both letters are held in the Hench collection.

Carroll's impression of the mosquito hypothesis—that it was

useless—come from his own words in a letter to an editor on June 26, 1903.

I pieced together the scene of Carroll's first symptoms of the illness from a few different sources. The description of sea bathing came from a letter that Lazear wrote to his mother on September 18, 1900, describing the water as feeling as warm as the air. Dr. William Bean's book, *Walter Reed,* also offers Alva Sherman Pinto's recollection of that afternoon.

The details about Carroll's illness were taken from Agramonte's account, Lena Warner's personal account of nursing Carroll and Howard A. Kelly's book.

The scene in which William E. Dean is infected has been debated over the years. In some accounts, including a popular play called *Yellow Jack,* Dean was infected knowingly or unknowingly while he was bedridden and recovering in Las Animas Hospital. For this book, I based the scene on Agramonte's own account, as he was one of the four members of the board.

Reed's letters to Kean during Carroll's illness, as well as his letter to Carroll on September 7, 1900, are held in the Hench collection.

Lazear's letter to his wife, Mabel, was written on September 8, 1900, and is held in the Hench collection. I refer to the debate over the resulting tragedy as one that continued through the next five decades because a number of historians and participants attempted to explain what happened. Philip S. Hench was still piecing the story together in the 1940s and 1950s—five decades after the incident.

Guinea Pig No. 1

The description of the hospital room at Las Animas is based on a photograph held in the Hench collection. For a time, that room was marked with a plaque in honor of Jesse Lazear. The information about *Aedes aegypti* as a vector came from Robert Desowitz's *Mosquito.*

The scene in this hospital room is a re-creation from Agramonte's *The Inside History of a Great Medical Discovery.* This was the story Jesse Lazear adamantly told colleagues—he never wavered from this account. His colleagues agreed that the story did not seem reasonable—Lazear would have known exactly what kind of mosquito landed on his arm, and he was far too meticulous to have let it go at that. Reed, Agramonte and Carroll all believed Lazear himself was the "Guinea Pig" in his logbook. Nonetheless, they assumed Lazear had his reasons for not telling the truth—reasons that even today are a mystery—so they kept to the story Lazear himself had told just before he died.

All information and recordings from the logbook were taken from the book itself on my visit to the New York Academy of Medicine.

Details about how Lazear spent his time—sea bathing and reading each night before bed—came from a letter he wrote to his mother on September 18, 1900. The quote about how much he missed Houston is also taken from that letter.

Lazear's complaint about feeling "out of sorts" came from Agramonte's account, and the description of how Lazear spent that first night with yellow fever—organizing his notes—was taken from Truby's *Memoir of Walter Reed.* In that book, Truby also describes the following morning when Lazear was taken by litter out of his home and moved into the yellow fever ward.

There are several references in Truby's writings and Gustav Lambert's account to Lena Warner nursing both James Carroll and Jesse Lazear. Warner's own account is full of inaccuracies, even untruths. Whenever taking facts directly from her account, I was sure to find a second or third source to back up her claims. I used creative license in the section where Warner remembers her own case of yellow fever. Her writings refer to how the incident stayed with her, so it seems natural to assume nursing fever patients took her back to that place and time.

The description of the record book required by the chief surgeon is based on my visit to the National Library of Medicine where the original record books can be found. All that is left of Jesse Lazear's is his fever chart, which is part of the Hench collection.

James Carroll's remark about being profoundly shaken by the sight of his friend came from his interview with Caroline Latimer, published in *A Cyclopedia of American Medical Biography*. Agramonte's impression was found in his *The Inside History of a Great Medical Discovery*.

Walter Reed's letter to Carroll was written on September 24, 1900, and Reed's letter to Jefferson Randolph Kean was written on September 25, 1900. Both letters are held in the Hench collection.

The description of Lazear's spiraling illness and eventual death were based on Lena Warner's "Recollections of Lena A. Warner." Similar details were taken from Gustav Lambert's account, Truby's memoir and Hench's research. His fever chart shows his temperature falling from 104 to 99 degrees, where it flatlined. In trying to do justice to Lazear's horrible death, I relied wholly on facts recorded by others—his running madly around the room destroying things and the vomit roiling over the bar. In that instance, it is not bars of the cot but the mosquito bar and netting hanging over the hospital bed. The only point when I added a detail not explicitly described firsthand was in restraining Lazear. Warner recalled two soldiers having to hold him down and restrain him, but there is no record of *how* they restrained him. In this account, I presumed they tied his wrists and ankles.

A copy of Jesse Lazear's death certificate can be found in the Hench collection, and the original is at the National Library of Medicine. The account of his burial was based on Truby's description. There have been discrepancies about whether or not James Carroll was present at the funeral. Reed was in the United States, and Agramonte had just been sent there on orders from General Wood (copies of those orders are in the Hench collection). Some sources have excluded Carroll's presence or said that no members of the board were present; however, James Carroll wrote to his wife, Jennie, on September 28, 1900, that he had just returned from Lazear's burial.

Details of how Mabel Lazear learned of her husband's death come from Hench's "An Illustrated Talk by Philip S. Hench" on January 31, 1955, as well as Hench's "Interview with Jefferson Randolph Kean," on January 6, 1944. Mabel's letter to Carroll, dated November 10, 1900, is part of the Hench collection.

The account of Reed retrieving Lazear's logbook is based on Truby's account, as well as Bean's *Walter Reed*.

Camp Lazear

Details of Walter Reed's return to Cuba aboard the *Crook* were part of Hench's "The Conquest of Yellow Fever." The account of Robert P. Cooke sharing a cabin with Reed comes from *Yellow Jack*. The letter reprimanding Cooke, written on July 24, 1900, was written by the acting chief surgeon, Alexander Stark. The letter is part of the Hench collection.

Statistics about the yellow fever epidemic in Havana that year came from Bean's book.

Reed's general depression over the death of Lazear was noticed by Truby, as well as others at Camp Columbia. He also wrote to Emilie about it. His guilt at being in the United States while his board self-experimented was recorded in his letter to Kean on September 25, 1900.

Truby's *Memoir of Walter Reed* describes the following weeks when Reed wrote and researched his paper on yellow fever. He also related the scene when Reed questioned William Dean about his yellow fever case. Reed's paper "The Etiology of Yellow Fever: A Preliminary Note" can be found in the Hench collection and at the National Library of Medicine. The excerpt from the *Indianapolis Journal* was taken from a letter by Mary Fishback to Philip S. Hench, August 30, 1940. The *New York Times* quote about the presentation appeared in their "Topics of the Times" on November 10, 1900. The criticism from the *Washington Post* was published on November 2, 1900.

The letter in which Sternberg informs Reed that he has submitted the paper for publication was dated October 23, 1900, and is part of the Hench collection. That paper appeared in the *Philadelphia Medical Journal* on October 27.

The account of Reed meeting with General Wood in the Governor's Palace, Havana, to request money for Camp Lazear was written in "A Review of Dr. Howard A. Kelly's Book *Walter Reed and Yellow Fever*," by Kean. The review was never published, but is held in the Hench collection.

Details about the development of Camp Lazear came primarily from Agramonte's account. Carroll later denied that Agramonte had anything to do with selecting the site for the camp, but it made the most sense to have Agramonte scout out a location. He had lived in Cuba the longest, and the final location of the camp was on a farm belonging to some of his friends—Finca San Jose in Marianao outside of Havana.

The description of LaRoche's books was based on personal observation. I looked through a copy of the original 1853 publication at the library of Southwestern Medical School in Dallas,

Texas. And the account of Reed quoting LaRoche, the storm that destroyed their batch of mosquitoes and the hunt for new ones, all came from Truby.

The dimensions and details about Building No. 1 (the Infected Clothing Building) and Building No. 2 (Infected Mosquito Building) came from a number of sources. First and foremost, in 2005, I visited the site of what remains of Camp Lazear in Marianao, which is now a slum section outside of Havana. Only Building No. 1 still stands, but it was discovered by Hench and John Moran in the 1940s and returned to its original state as part of a memorial park dedicated to the yellow fever experiments. Hench worked with a number of medical officials and the Cuban government under Batista to renovate the building and erect a memorial wall. By the time I visited in 2005, few people in Havana knew where the park was, and the building was in a state of disrepair. However, it was still the same dimensions that Reed designed, and looking through broken boards I could see remains of the original tongue and groove construction, although the wood was rotting in a number of places and patched together in others.

Other descriptions of the building came from Reed's hand-drawn plans, Truby's account and "Memorandum on Yellow Fever Experiments," written by Robert P. Cooke in 1940. Additional details about the men's experience inside the building came from Agramonte's article.

The majority of information about John Moran, how he came to be one of the volunteers, his stay in the Infected Mosquito Building and his resulting illness came from his own unpublished autobiography *Memoirs of a Human Guinea Pig,* written at Hench's request in 1937.

The story of Major Peterson was told in an account written by Kean on May 8, 1941, the *Recollections of Lena A. Warner* and *Recollections of Personal Experiences in Connection with Yellow Fever* by

Chauncey Baker. All three sources can be found in the Hench collection.

Information about Roger Post Ames came from Moran's account, as well as the account of Gustav Lambert, Ames's nurse.

Lawrence Reed's remark about the veracity of Reed's famous statement to Moran and Kissinger came from an interview with Lawrence and Blossom Reed by Hench on November 21, 1946.

Reed's written recommendation for John Moran was found in Moran's *My Date with Walter Reed and Yellow Jack*. Both Reed and Moran were on the same transport back to the U.S. Reed approached Moran and gave him the recommendation, adding that he should have done it long before then.

The account of procuring the Spanish volunteers at the Immigration Station came from Agramonte. And the details about the consent form came from Bean's book, *Walter Reed*.

The description of the experiments performed on John Kissinger came from his own account given to Hench, "Memorandum: Experiences with the Yellow Fever Commission in Cuba 1900, by John R. Kissinger." The details about how the men responded to Kissinger as a hero from that point forward was found in an account written by Paul L. Tate on July 27, 1954, for Hench. Tate also provided the "old army saying."

For the sake of pacing, I streamlined some of the details about the experiments in November, December and January. The facts are all there, but I chose not to outline every single experiment performed during that three-month period. To recreate the experiments, I relied on several different sources. Philip S. Hench provided a "Summary of Research," written on August 20, 1940, which I used as a general guideline. I also relied on the personal accounts written by Kissinger, Moran and Truby. Other details came out in original letters: One was written by Hench to Truby, January 7, 1941, and the other was a letter Reed wrote to Truby on

December 10, 1900. Reed's quote about the importance of these discoveries came from his letter to Emilie on December 9, 1900. An excerpt of that letter was provided by Blossom Reed in her "Biographical Sketch," written for Hench.

The rumors about the bleached bones of Walter Reed's yellow fever volunteers came from Bean's book, as well as a letter Reed wrote to Emilie.

In developing the scene where Walter Reed walks through the streets of Havana on his way to a banquet on December 22, 1900, I used personal experience, old photographs and Hench's notes. It was December when I visited Havana, so I had the opportunity to see huge poinsettia bushes in bloom and Christmas decorations around the city. I visited Parque Central, the Hotel Inglaterra and the site of what used to be Old Delmonico's. In Hotel Inglaterra, they have an old print of the area from 1904. From the sketch, I took details of the park as it appeared in 1900, as well as the Tacón theater, which is now the beautiful Gran Teatro. The building where the restaurant stood in 1900 is now abandoned, but I was still able to climb the staircase and study details about the architecture. Details about the banquet itself came from a speech given by Hench on December 3, 1952, entitled, "The Historic Role of the Finca San Jose and Camp Lazear in the Conquest of Yellow Fever by Carlos Finlay, Walter Reed and Their Associates." Information about Finlay's career after the Reed experiments—the fact that he was nominated for the Nobel Prize seven different times—came from the official website for the Nobel Prize, which lists past nominees and winners.

Carroll's letter to Jennie is held in the Carroll Box, as part of the Hench collection at the University of Virginia. The description of the Christmas party, the makeshift mosquito and the poem for Reed came from letters he wrote to Emilie on December 25 and 26, 1900.

A New Century

Walter Reed's original letter to Emilie on New Year's Eve is part of the Philip S. Hench Walter Reed Collection at the University of Virginia.

Blood

Details surrounding the blood inoculations came from two main sources: Truby provided some background information in his memoir, but the majority of the chapter came from John H. Andrus's "I Became a Guinea Pig," held in the Hench collection.

Instructions from the surgeon general came from letters exchanged between Sternberg and Reed in December 1900.

The description of Roger Post Ames was taken from Lambert's account, as well as Paul Tate's "Essay: Roger Post Ames," written for Hench in 1954.

Reed's letter to Sternberg expressing his concern for Andrus was written on January 31, 1901. Excerpts from that letter appear in Andrus's own account, as well as the biography *George M. Sternberg*.

The Etiology of Yellow Fever

The account of Reed's presentation to the Pan-American Medical Congress in Havana came from his own descriptions in family letters submitted to the Philip S. Hench collection by Blossom Reed.

The remark about Reed's voice rising to a falsetto note when he emphasized important points was taken from Hench's interview with Lawrence Reed on November 21, 1946. The quote about Reed as a teacher came from Captain J. Hamilton Stone's remarks in Kelly's book *Walter Reed*.

The *Washington Post* quote was from a clipping dated February 11, 1901, held in the Hench collection.

Retribution

Walter Reed was given military orders to report to Buffalo, New York, on September 5, 1901. Those original orders are held in the Philip S. Hench Walter Reed Collection.

The account of McKinley's assassination came from a *New York Times* article published on September 7, 1901. The facts of the article were also checked against *Encyclopaedia Britannica.*

Proceedings from the 1901 American Public Health Association meeting are held in the Hench collection under the title *Public Health Papers and Reports, Volume XXVII, Presented at the Twenty-ninth Annual Meeting of the American Public Health Association, Buffalo, NY, September 16–20, 1901.*

The information about Wasdin's theory of a poisoned bullet came from a *New York Times* article dated September 16, 1901.

Reed's opinion of the Guitéras experiments came from a letter written by Reed to Gorgas, May 23, 1901. The letter is held in the Walter Reed Papers at the National Library of Medicine. Reed's disappointment in hearing about the deaths resulting from the Guitéras experiments was found in another letter to Gorgas, dated September 2, 1901, also held at the National Library of Medicine.

The account of Clara Maass's death during the experiments came from Philip S. Hench's personal notes, as well as a *New York Times* article published on August 25, 1901. James Carroll's experiments passing blood through the Berkefeld filter were outlined in his *Report to the Surgeon General,* August 18, 1906, held in the Hench collection.

Reed's frustration in being passed over for surgeon general, as well as his statements about doing something for the real benefit of humanity, were expressed in a letter to Gorgas on July 21, 1902. The letter is part of the Walter Reed Papers at the National Library of Medicine.

The description of Keewaydin and the inscription above the fireplace was relayed in a "Biographical Sketch of Walter Reed" written by Emilie Lawrence Reed, held in the Hench collection.

The quote about Reed's failing health and his frustration over having persons in high authority rob him of his just fame was taken from a letter from Henry Hurd, a friend of Reed's, to Caroline Latimer on February 11, 1905. The letter can be found in the Hench collection. A similar reference, though with slightly different wording, can be found in Kelly's book.

The statements by George M. Sternberg appeared in the July 1901 *Popular Science Monthly*. Reed's anger at these remarks was expressed in a letter to Gorgas dated July 27, 1901, and held at the National Library of Medicine. Sternberg's letter requesting a promotion to major general was dated January 25, 1901, and is held in the Sternberg papers at the National Library of Medicine.

The description of Reed's final days and illness came from Emilie Lawrence Reed in notes held in the Hench collection, as well as "Notes on Reed and Carroll," written by Philip S. Hench on January 10, 1942. Additional details were found in the Walter Reed Papers at the National Library of Medicine. The account of Reed's death was taken from three sources: Kean's letter to Howard Kelly on March 25, 1901, William Borden's letter to Howard Kelly on March 16, 1905, and the *Report: History of Doctor Walter Reed's Illness from Appendicitis* by William Borden, 1903. All three are held in the National Library of Medicine.

Details of Reed's funeral were taken from Kean's letter to Howard Kelly on March 25, 1901, Truby's recollection, Hench's interview with Lawrence and Blossom Reed on November 21, 1946, and a *Biographical Sketch: Life and Letters of Dr. Walter Reed by His*

Daughter. Welch's remarks were found in the *Message from the President of the United States Transmitting Certain Papers in regard to Experiments Conducted for the Purpose of Coping with Yellow Fever,* by Theodore Roosevelt, December 5, 1906. Roosevelt's quote about Reed's contribution to the betterment of mankind came from Senate Document No. 10, Fifty-ninth Congress. All of the above are held in the Hench collection.

The list of names who contributed to the Walter Reed Memorial Association were found in Writer and Pierce's *Yellow Jack* and Bean's *Walter Reed.*

The Mosquito

Biographical information about Major William C. Gorgas came from Greer Williams's book *The Plague Hunters,* as well as *William Crawford Gorgas: His Life and Work.*

In an interview with an anonymous source in Havana, I confirmed that the same method used in 1900 for monitoring mosquitoes is still in use today.

Part IV: United States, Present Day

Epidemic

The introductory quote for Part IV was found in T. P. Monath's "Yellow Fever: An Update," published in *Lancet Infectious Disease,* 2001.

The account of Tom McCullough's death from yellow fever came from the Centers for Disease Control and Prevention: "Fatal Yellow Fever in a Traveler Returning from Amazonas, Brazil, 2002," *Morbidity and Mortality Weekly Report.* I also consulted another article, "Fatal Yellow Fever in a Traveler Returning from

Venezuela, 1999," in the *Morbidity and Mortality Weekly Report* for information about similar cases. Some of the more personal details about McCullough's hospital stay and death were taken from two articles in the Corpus Christi *Caller-Times,* March 27, 2002, and May 14, 2004.

To describe what would happen in the case of an epidemic in the United States, I followed the CDC's "Response to an Epidemic of Yellow Fever," published in November 2005 specifically for Africa and the Americas. In the report, the CDC outlines the response of field investigators, armed forces, border officials, medical personnel, educational campaigns and vaccine usage.

For additional information about yellow fever vaccine production and stockpiling, I consulted the WHO's "State of the World's Vaccines and Immunizations" and the Global Alliance for Vaccines and Immunization. Prior to 2002, there was a global shortage of the yellow fever vaccine due to the lack of funds and too few labs producing the vaccine. Since then, the GAVI, with help from the Vaccine Fund, has been able to keep stockpiles of six million vaccines in the case of an epidemic, as well as an additional six million for yearly routine use in African and South American countries where yellow fever is endemic. According to the WHO report, there are four main manufacturers of the yellow fever vaccine, with a total global production capacity of 270 million.

A Return to Africa

Character development of Adrian Stokes came primarily from Greer Williams's *The Plague Killers,* written in 1969. Some additional information about Stokes, as well as some of the details about the Rockefeller compound in Yaba, were found in Charles Bryan's *A Most Satisfactory Man: The Story of Theodore Brevard Hayne, Last Martyr of Yellow Fever.*

Descriptions of the Rockefeller Foundation were gathered from Williams's book and the website for the Rockefeller Foundation. John M. Barry's *Influenza* also provided some material about the historical significance of the Rockefeller Institute and Rockefeller Foundation.

The two quotes cited in this chapter were taken from Laurie Garrett's *The Coming Plague,* written in 1994, and Paul De Kruif's *Microbe Hunters,* written in 1926.

The Vaccine

Biographical information about Max Theiler and his work with the 17-D yellow fever vaccine came from Greer Williams's books *The Virus Hunters* and *The Plague Killers.* Williams was a contemporary of Theiler and was able to interview him personally for his book.

History Repeats Itself

The majority of updated information about yellow fever was taken from the World Health Organization.

Additional information about the attempts to eradicate *Aedes aegypti* from the United States was found in Andrew Spielman and Michael D'Antonio's book *Mosquito.* The quote about America going to war with Spain, in part, because of yellow fever was taken from their book.

Information about the Asian tiger mosquito and its discovery in Memphis came from Paul Reiter and Richard Darsie's "Aedes Albopictus in Memphis, Tennessee (USA): An Achievement of Modern Transportation," published in *Mosquito News,* 1984. Reiter was the entomologist who found the tiger mosquito in Memphis, TN. Additional details came from Gary Taubes's "Tales of a

Bloodsucker—Asian Tiger Mosquitoes," published in *Discover*, July 1998.

The recent study about the proteins on the surface of the yellow fever virus was published in an article in *Virology*, July 5, 2005. The study of the way a flavivirus interacts with interferon during an immune response was published in the *Journal of Virology*, September 2005.

The quote regarding *A. aegypti* mosquitoes established in urban areas was taken from the article "Yellow Fever: A Decade of Re-emergence," by S. E. Robertson, et al, the *Journal of the American Medical Association*, 1996.

Epilogue: Elmwood

The majority of the descriptions of Elmwood were based on several visits there to look through their historical collections and an interview with superintendent Sunny Handback just before he retired in November 2005. The reference to the terms *burial* and *cemetery* were taken from the book *Elmwood: In the Shadow of Elms*. I also read through Elmwood's ledger of burials for 1878–1879.

Selected Bibliography

Archives and Collections

American Lloyd's Register of American and Foreign Shipping 1865, "Emily B. Souder."

Claude Moore Health Sciences Library, University of Virginia
   The Jefferson Randolph Kean Papers
   The Philip S. Hench Walter Reed Yellow Fever Collection
   Public Health Papers and Reports, presented at the Twenty-Ninth Annual Meeting of the American Public Health Association, Buffalo, NY, September 16–20, 1901
   Senate Document No. 822
   The Wade Hampton Frost Papers
   Walter Reed Letters
   The William Bennett Bean Papers

Dee J. Canale, M.D., Yellow Fever and Medical History Private Collection

Elmwood Cemetery
   Charles C. Parsons File
   Ledger for August and September 1878 burials at Elmwood
   William J. Armstrong, Armstrong Family File

Health Sciences Historical Collection, University of Tennessee Library of Medicine
   Simon R. Bruesch Collection

Library of Congress, Rare Books Collection
   Conclusions of the board of experts authorized by Congress to investigate the yellow fever epidemic of 1878: being in reply to questions of the committees of the Senate and House of Representatives of the Congress of the United States, upon the subject of epidemic diseases. Washington, DC: 1879.
   Proceedings of the Board of Experts authorized by Congress, to investigate the yellow fever epidemic of 1878: Meeting held in Memphis, Tenn., December 26th, 27th, 28th, 1878. Washington, DC: 1879.
   Sigsbee, Charles Dwight. *The Maine.* New York: The Century Co., 1899.
   "Yellow Fever Bill." Washington, DC: 1879.

Memphis History Exhibit, Pink Palace Museum

Mississippi Valley Collection, University of Memphis
   Caleb Goldsmith Forshey Diaries
   Charles G. Fisher Papers
   De La Hunt Papers
   Eldon Roark Papers
   Hughetta Snowden Papers
   Jefferson Davis Papers
   Mary Louise Costillo Nichols Scrapbook

Pinch District Collection
Porter-Rice Family Papers
U.S. Department of Agriculture Weather Bureau, Memphis Station Records, 1878

National Archives and Records Administration
Records of Public Buildings Service
Record Group 112

National Library of Medicine, History of Medicine Collection
Albert Ernest Truby Papers (1898–1953)
Fever Epidemic at Columbia Barracks Collection
George Miller Sternberg Papers (1861–1912)
Walter Reed Papers (1898–1902)

New York Academy of Medicine
"Record of the Yellow Fever Commission's Work." Archibald Malloch Collection.

Record of American and Foreign Shipping 1871, "Emily B. Souder."

Yellow Fever Collection, Memphis Library
Charles Carroll Parsons Papers
General Colton Greene File
George C. Harris Papers
Howard Association Collection
John H. Erskine File
John Ogden Carley Papers
Lena A. Warner File
Louis Schuyler Papers
Summary of Minutes of Board of Health, City of Memphis, 1870–1905
William J. Armstrong Papers

Books and Articles

Agramonte, Aristides. *The Inside History of a Great Medical Discovery.* Havana: Times of Cuba Press, 1915.

Altman, Lawrence K., M.D. *Who Goes First?* Berkeley: University of California Press, 1986.

Anderson, Laurie Halse. *Fever 1793.* New York: Aladdin, 2002.

Baker, Christopher. *Cuba.* Third Edition. Emeryville, CA: Avalon Travel Publishing, 2004.

Baker, Thomas. "Yellowjack: The Yellow Fever Epidemic of 1878 in Memphis, Tennessee." *Bulletin of the History of Medicine,* Vol. 42, No. 3 (1968).

Barry, John M. *The Great Influenza: The Epic Story of the Deadliest Plague in History.* New York: Viking Penguin, 2004.

Bean, William B., M.D. *Walter Reed, A Biography.* Charlottesville: University Press of Virginia, 1982.

Bemiss, S. M. "Report upon Yellow Fever in Louisiana in 1878." *New Orleans Medical and Surgical Journal,* n.s., XI (1883): 82–86.

Best, S., et al. "Inhibition of interferon-stimulated JAK-STAT Signaling by tick-borne Flavivirus of NS5 as interferon antagonist." *Journal of Virology* (Sept. 2005).

Biennial Report—Memphis Board of President of Fire and Police Commissioners of the Taxing District (Memphis), Shelby County, Tennessee, to the Governor of the State. December 1, 1880.

Bloom, Khaled J. *The Mississippi Valley's Great Yellow Fever Epidemic of 1878.* Baton Rouge and London: Louisiana State University Press, 1993.

Bond, Beverly G., and Janann Sherman. *Memphis: In Black and White.* Chicago: Arcadia Publishing, 2003.

Brands, H. W. *The Reckless Decade: America in the 1890s.* New York: St. Martin's Press, 1995.

Bray, R. S. *Armies of Pestilence: The Impact of Disease on History.* New York: Barnes and Noble Books, 1996.

Bristow, Eugene. "From Temple to Barn: The Greenlaw Opera House in Memphis, 1860–1880." *West Tennessee Historical Society Papers,* XXI, 1967.

Bruesch, Simon Rulin, M.D. "The Disasters and Epidemics of a River Town: Memphis, Tennessee, 1819–1879." Reprinted from *Bulletin of the Medical Library Association,* Vol. 40, No. 3 (July 1952).

Bruesch, Simon Rulin, M.D. "Yellow Fever in Tennessee in 1878." *Journal of the Tennessee Medical Association,* Part I (December 1978), Part II (February 1979), Part III (March 1979).

Bunnell, Joseph. "Killer Virus." *University of Texas Medical Branch at Galveston Quarterly.* Winter 2001: 16–19.

Burnside, Madeleine, and Rosemarie Robotham. *Spirits of the Passage: The Transatlantic Slave Trade of the Seventeenth Century.* New York: Simon & Schuster, 1997.

Buser, Lawrence. "City Still Bears Scars of Epidemic Century Ago." *The Commercial Appeal,* June 18, 1978.

Bynum, W. F. *Science and the Practice of Medicine in the Nineteenth Century.* New York: Cambridge University Press, 1994.

Capers, Gerald M., Jr. *The Biography of a River Town, Memphis: Its Heroic Age.* New Orleans: Tulane University, Published by Gerald M. Capers, 1966.

Carrigan, Jo Ann. *The Saffron Scourge: A History of Yellow Fever in Louisiana.* Lafayette: University of Louisiana Press, 1994.

Carrigan, Jo Ann. "Yellow Fever: Scourge of the South." *Disease and Distinctiveness in the American South.* Knoxville: The University of Tennessee Press, 1988.

Carroll, J. "A Brief Review of the Aetiology of Yellow Fever." *New York Medical Journal and Philadelphia Medical Journal* 79 (1904): 241–45, 307–10.

Carter, Henry Rose. *Yellow Fever: An Epidemiological and Historical Study of Its Place of Origin* (1931).

Centers for Disease Control and Prevention. "Fatal Yellow Fever in a Traveler Returning from Amazonas, Brazil, 2002." *Morbidity and Mortality Weekly Report* (April 19, 2002): 324–25.

Centers for Disease Control and Prevention. "Fatal Yellow Fever in a

Traveler Returning from Venezuela, 1999." *Morbidity and Mortality Weekly Report* (April 14, 2000): 303–5.

Centers for Disease Control and Prevention. "Response to an Epidemic of Yellow Fever." (November 3, 2005).

Choppin, Samuel. "History of the Importation of Yellow Fever into the United States, 1693–1878." *Public Health Papers, American Public Health Association,* Vol. 4 (1877–1878).

Cloudesley-Thompson, J. L. *Insects and History.* London: Weidenfeld & Nicholson, 1976.

Coleman, William. *Yellow Fever in the North: The Methods of Early Epidemiology.* Madison: University of Wisconsin Press, 1987.

Community of St. Mary, the Sisters of St. Mary at Memphis: With the Acts and Sufferings of the Priests and Others Who Were There With Them During the Yellow Fever Season of 1878. New York, 1879.

Connell, Mary Ann Strong. "The First Peabody Hotel: 1869–1923." *West Tennessee Historical Society Papers,* Vol. 28–30 (1974): 76.

Constance and Her Companions, the Martyrs of Memphis.

Coppock, Helen, and Charles Crawford. *Paul Coppock's Midsouth,* Vol. III (1976–1978). Nashville: Williams Printing, 1993.

Coppock, Paul. *Memphis Memoirs.* Memphis: Memphis State University Press, 1980.

Coppock, Paul. *Memphis Sketches.* Memphis: Friends of Memphis & Shelby County Libraries, 1976.

Coppock, Paul. "Memphis' No. 1 Fighter of Yellow Fever." *The Commercial Appeal,* June 9, 1974.

Costillo, Mary L. "Reminiscences of My Childhood and Youth." *West Tennessee Historical Papers* 12 (1958): 80–1081.

Crane, Stephen. *This Majestic Lie.* (1900).

Crawford, Charles W. *Yesterday's Memphis.* Miami, FL: E. A. Seemann Publishing, 1976.

Cushing, Harvey. *The Life of Sir William Osler.* London: Oxford University Press, 1940.

Dando, Mary. "Our Immigrant Heritage: The Irish in Memphis." *Memphis Magazine* (September 2003).

Davis, J. H. *St. Mary's Cathedral 1858–1958*. Memphis: Published by the Chapter of St. Mary's Cathedral, 1958.

Davis, J. H. "Two Martyrs of the Yellow Fever Epidemic of 1878." *West Tennessee History Society Papers*, Vol. 26 (1972): 20–39.

Davis, James. *The History of the City of Memphis*. Memphis: Hite, Crumpton & Kelly Printers, 1873.

De Kruif, Paul. *Microbe Hunters*. New York: Harcourt, 1926.

Delaporte, F. *History of Yellow Fever: An Essay on the Birth of Tropical Medicine*. Cambridge, MA: MIT Press, 1991.

Del Regato, J. A. "Jesse William Lazear: The Successful Experimental Transmission of Yellow Fever by the Mosquito." *Medical Heritage*, Vol. 2, No. 6 (November–December 1986).

Desowitz, Robert S. *The Malaria Capers: Tales of Parasites and People*. New York: W. W. Norton, 1991.

Desowitz, Robert S. *Who Gave Pinta to the Santa Maria?* New York: W. W. Norton, 1997.

Diamond, Jared. *Guns, Germs, and Steel: The Fates of Human Societies*. New York: W. W. Norton, 1999.

Diaz, Henry F., and Gregory J. McCabe. "A Possible Connection between the 1878 Yellow Fever Epidemic in the Southern United States and the 1877–78 El Niño Episode." *Bulletin of the American Meteorological Society*, September 30, 1998: 21–28.

Dromgoogle, Dr. J. P. *Yellow Fever Heroes, Honors, and Horrors of 1878*. Louisville: John P. Morton, 1879.

Durham, Herbert and Walter Myers. "Yellow Fever Expedition." *British Medical Journal*, September 8, 1900.

Eaton, Tim. "Family of Yellow Fever Victim Loses Its Lawsuit." Corpus Christi *Caller-Times*, May 14, 2004.

Eckstein, Gustav. *Noguchi*. New York: Harper & Brothers, 1931.

Ellis, J. H. *Yellow Fever and Public Health in the New South*. Lexington: University of Kentucky Press, 1992.

Ellis, John H. "Disease and the Destiny of a City: The 1878 Yellow Fever Epidemic in Memphis." *West Tennessee Historical Society Papers* 28 (1974): 75–89.

Elmwood: History of the Cemetery. Memphis: Boyle and Chapman Printers, Publishers and Binders, 1874.

Erskine, John H. "A Report on Yellow Fever as It Appeared in Memphis, Tenn., in 1873." *American Public Health Association, Public Health Papers and Reports,* Vol. I (1873).

Finger, Michael. "The Martyrs of Memphis." *Memphis Magazine,* 1999.

Finger, Michael. "When Cotton Was King." *Memphis Magazine,* City Guide, 2003.

Finlay, Carlos E. *Carlos Finlay and Yellow Fever.* New York: Oxford University Press, 1940.

Fitch, S. S. *The Family Physician.* New York, 1876.

Fowinkle, Eugene, M.D., and Mildred Hicks. "Development of Public Health and the Yellow Fever Epidemics in Memphis." *History of Medicine in Memphis.* Jackson, TN: McCowat-Mercer Press, 1971.

"Fragment of YFV May Hold Key to Safer Vaccine," *Medical News Today* (July 17, 2005).

Garrett, Laurie. *The Coming Plague: Newly Emerging Diseases in a World Out of Balance.* New York: Penguin Books, 1994.

"George Waring Obituary." *The New York Times,* October, 30, 1898.

Gillett, Mary. "A Tale of Two Surgeons." *Medical Heritage,* November/December 1985.

Goddard, J. *Physician's Guide to Arthropods of Medical Importance.* Boca Raton, FL: CRC Press, 2003.

Goodman, Dr. Louis, and Dr. Alfred Gillman. *The Pharmacological Basis of Therapeutics.* Second Edition. New York: Macmillan, 1955.

Gorgas, Marie. *William Crawford Gorgas: His Life and Work.* New York: Doubleday, 1924.

Gorgas, W. C. "Sanitation of the Tropics with Special Reference to Malaria and Yellow Fever." *Journal of the American Medical Association* 52 (1909): 1075–77.

Gorn, Elliott J. *Mother Jones.* New York: Hill and Wang, 2001.

Gould, Lewis L. *America in the Progressive Era, 1890–1914.* New York: Longman, 2001.

Greenhill, E. Diane, R.N., B.S.N., Ed.D. *From Diploma to Doctorate: 100 Years of Nursing.* Memphis: University of Tennessee Press, 1988.

Groh, Lynn. *Walter Reed, Pioneer in Medicine.* New York: Dell Publishing, 1971.

Guitéras, Juan. "Experimental Yellow Fever at the Inoculation Station of the Sanitary Department of Havana with a View to Producing Immunization." *American Medicine,* November 23, 1901.

Halle, Arthur. "History of the Memphis Cotton Carnival." *West Tennessee Historical Society Papers,* Vol. I (1952).

Harkins, John E. *Metropolis of the American Nile, Memphis and Shelby County.* Oxford, MS: The Guild Bindery Press, 1982.

Harris, George C. "Memorial Sermon Preached in St. Mary's Cathedral, Memphis, December 22, 1878." New York: 1878.

Hatcher, J. Edward, Jr. *Gayoso Bayou.* Memphis: St. Luke's Press, 1982.

Hemmeter, John C. "Major James Carroll of the United States Army, Yellow Fever Commission, and the Discovery of the Transmission of Yellow Fever by the Bite of the Mosquito 'Stegomyia Fasciata.'" *American Public Health Reports,* 1908.

Hicks, M. (ed.). *Yellow Fever and the Board of Health, Memphis, 1878.* The Memphis and Shelby County Health Department, 1964.

Higman, B. W. *Slave Populations of the British Caribbean, 1807–1834.* Baltimore: Johns Hopkins University Press, 1988.

"Horrors of Plague Live on Thru Years." *The Evening Appeal,* December 27, 1932.

Howard, Leland Ossian. *Mosquitoes: How They Live; How They Carry Disease; How They Are Classified; How They May Be Destroyed,* 1901.

Hume, Edgar Erskine. "Sternberg's Centenary, 1838–1938." *The Military Surgeon* 84 (1939): 420–28.

Humphreys, Margaret. *Yellow Fever and the South.* Baltimore: The Johns Hopkins University Press, 1992.

"Hydropathy Used in Fever Epidemic." *The Night Desk, The Commercial Appeal,* January 16, 1954.

"Incidents of the Scourge at the South." *Frank Leslie's Illustrated Newspaper,* September 21, 1878.

Keating, J. M. *The Yellow Fever Epidemic of 1878, in Memphis, Tennessee.* Memphis: Printed for the Howard Association, 1879.

Kelly, Howard A. *Walter Reed and Yellow Fever.* New York: McClure, Phillips, 1906.

Kolata, Gina. *Flu: The Story of the Great Influenza Pandemic of 1918 and the Search for the Virus That Caused It.* New York: Farrar, Straus and Giroux, 1999.

Lanier, Robert. "Memphis Greets War With Spain." *The West Tennessee Historical Society Papers,* No. 18 (1964).

LaPointe, Patricia M. *From Saddlebags to Science: A Century of Health Care in Memphis, 1830–1930.* Memphis: Health Sciences Museum Foundation, 1984.

La Roche, R. *Yellow Fever.* Philadelphia: Blanchard and Lea, 1855.

Latimer, C. W. "James Carroll." in H. A. Kelly and W. L. Burrage, *A Cyclopedia of American Medical Biography.* Baltimore: Norman, Remington, 1920.

Lederer, Susan E. *Subjected to Science, Human Experimentation in America Before the Second World War.* Baltimore: The Johns Hopkins University Press, 1995.

"Lena Warner Obituary." *The Commercial Appeal,* August 19–20, 1948.

Leonard, Jonathan. "Carlos Finlay's Life and the Death of Yellow Jack." *Bulletin of The Pan-American Health Organization* 23–24 (1989): 438–52.

"Loss of the *Emily B. Souder.*" *The New York Times,* January 17, 1879.

Mackie, Dr. Thomas, Dr. George Hunter, and Dr. C. Worth. *A Manual of Tropical Medicine.* Second Edition. Philadelphia and London: W. B. Saunders Company, 1955.

Maegraith, B. G. "History of the Liverpool School of Tropical Medicine." *Medical History,* Vol. 16, No. 4 (1972): 354–68.

Magness, Perre. *Elmwood: In the Shadow of Elms.* Published by Elmwood Cemetery, 2001.

Magness, Perre. *Past Times: Stories of Early Memphis.* Memphis: Mercury Printing, 1994.

Malkin, Harold M. "The Trials and Tribulations of George Miller

Sternberg (1838–1915)—America's First Bacteriologist." *Perspectives in Biology and Medicine* 36.4 (Summer 1993): 666–78.

Millard, Candice. *River of Doubt: Theodore Roosevelt's Darkest Journey.* New York: Doubleday, 2005.

Monath, T. P. "The 1970 Yellow Fever Epidemic in Okwoga District, Benue Plateau State, Nigeria. 2: Epidemiological Observations." *Bulletin of the World Health Organization* 49 (1973).

Monath, T. P. "The 1970 Yellow Fever Epidemic in Okwoga District, Benue Plateau State, Nigeria. 2: Immunity Survey to Determine Geographic Limits and Origins of the Epidemic." *Bulletin of the World Health Organization* 49 (1973).

Monath, T. P. "Yellow Fever: An Update." *Lancet Infectious Diseases* 1 (2001): 11–20.

Monath, T. P. "Yellow Fever: Victor, Victoria? Conqueror, Conquest? Epidemics and Research in the Last Forty Years and Prospects for the Future." *American Journal of Tropical Medicine,* Vol. 45, No. 1 (1991).

Muñoz-Jordán, Jorge L., et al. "Inhibition of Alpha/Beta Interferon Signaling by the NS4B Protein of Flaviviruses." *Virology,* Vol. 79, No. 13 (2005).

Myers, Anna. *Graveyard Girl.* New York: Walker, 1995.

Nasidi, A., T. P. Monath, K. DeCock, et al. "Urban Yellow Fever Epidemic in Western Nigeria, 1987." *Transactions of the Royal Society for Tropical Medicine and Hygiene* 83 (1989): 401–6.

"News Feature: Globalization—How Healthy?" *Bulletin of the World Health Organization,* 79 (2001).

Norman, C. "The Unsung Hero of Yellow Fever?" *Science,* Vol. 223 (1984): 1370–72.

Oldstone, Michael B. A. *Viruses, Plagues, and History.* Oxford: Oxford University Press, 1998.

Ornelas-Struve, Carole and Joan Hassell. *Memphis, 1800–1900, Volume III: Years of Courage.* New York: Nancy Powers, 1982.

O'Toole, G.J.A. *The Spanish War: An American Epic 1898.* New York: W. W. Norton and Company, 1984.

"Pan-American Medical Conference." *Journal of the American Medical Association* 36: 461–62 and 446–47.

Peller, S. "Walter Reed, C. Finlay, and their Predecessors Around 1800." *Bulletin of the History of Medicine* 33 (1959): 195–211.

Petri, William A. "America in The World: 100 Years of Tropical Medicine and Hygiene." *American Journal of Tropical Medicine and Hygiene,* 71 (1), 2004.

Pierce, John R., and Jim Writer. *Yellow Jack.* Hoboken, NJ: John Wiley & Sons, 2005.

The Pinch, Market Square, Brinkley Park: Neighborhood Story and a Guide Map of Historical Places.

Plunkett, Kitty. *Memphis: A Pictorial History.* Norfolk, VA: The Donning Company, 1976.

Porteous, Clark. "So New York City Thinks It Has Problems, Ask Memphis About Yellow Fever Epidemic." *Press-Scimitar,* July 14, 1975.

Quinn, Rev. D. A. *Heroes and Heroines of Memphis or Reminiscences of the Yellow Fever Epidemics.* Providence, RI: E. L. Freeman & Son, 1887.

Reed, W., and J. Carroll. "The Etiology of Yellow Fever." *American Medicine* 3 (1902): 301.

Reiter, Paul and Richard Darsie. "Aedes albopictus in Memphis, Tennessee (USA): An Achievement of Modern Transportation." *Mosquito News* (1984).

Reiter, Paul. "Global Warming and Vector-Borne Disease: Is Warmer Sicker?" Competitive Enterprise Institute, July 28, 1998.

"Reported Loss of the Steam-Ship *Emily B. Souder.*" *The New York Times,* December 28, 1878.

"Resurgence of Yellow Fever." *World Health Forum* 14 (1993).

Riedel, Nora Huber, ed. and trans. *Yellow Fever Quarantine in Memphis, Tennessee, August 14–October 30, 1878.* Excerpts from the Diary of Henry Sieck, Pastor of Trinity Lutheran Church, Memphis, Tennessee.

Robertson, S. E., B. P. Hull, O. Tomori, O. Bele, J. LeDuc, and K. Esteves. "Yellow Fever: A Decade of Re-emergence." *Journal of the American Medical Association,* Vol. 276, No. 14 (1996): 1157–62.

Schlereth, Thomas J. *Victorian America: Transformations in Everyday Life*. New York: Harper Perennial, 1991.

Segel, Lawrence, M.D. "The Yellow Fever Plot: Germ Warfare during the Civil War." *The Canadian Journal of Diagnosis*, 2002.

Sigafoos, Robert A. *Cotton Row to Beale Street*. Memphis: Memphis State University Press, 1979.

The Sisters of St. Mary at Memphis: With the Acts and Sufferings of the Priests and Others Who Were There with Them during the Yellow Fever Season of 1878. New York: Printed, but not Published (1879). Transcribed by Elizabeth Boggs and Richard Mammana, 2000–2001.

Solorazano, Armando. "Sowing the Seeds of Neo-imperialism: The Rockefeller's Yellow Fever Campaign in Mexico." *International Journal of Health Services*, 1993.

Sorrels, William W. *Memphis' Greatest Debate; a Question of Water*. Memphis: Memphis State University Press, 1970.

Spielman, Andrew, and Michael D'Antonio. *Mosquito: The Story of Man's Deadliest Foe*. New York: Hyperion, 2001.

Starr, Paul. *The Social Transformation of American Medicine*: New York: Basic Books, 1982.

Sternberg, G. M. "The Address of the President." *Journal of the American Medical Association*, Vol. 30 (1898): 1373–80.

Sternberg, George M. "The Bacillus Icteroides (Sanarelli) and Bacillus X (Sternberg)." *Transactions of the Association of American Physicians* 13 (1898): 70–71 and discussion by William Osler: 61–72.

Sternberg, George M. *Yellow Fever*. Extracted from *The American System of Practical Medicine*. Philadelphia and New York: Lea Brothers, 1897–98.

Sternberg, George M., and Walter Reed. "Report on Immunity against Vaccination Conferred upon the Monkey by Use of the Serum of the Vaccinated Calf and Monkey." *Transactions of the Association of American Physicians* 10 (1895): 57–69.

Sternberg, Martha. *George Miller Sternberg: A Biography*. Chicago: American Medical Association, 1920.

Stewart, Walter. "Bring Out Your Dead, Cried Yellow Fever." *Press-Scimitar*, April 7, 1932.

Strong, Philip. "Epidemic Psychology: A Model." *Sociology of Health & Illness,* Vol. 12, No. 3 (1990).

Sullivan, M. *Our Times: The Turn of the Century.* New York: Charles Scribner's Sons, 1937.

Summers, Thomas O., M.D. *Yellow Fever.* Nashville: Wheeler Bros., 1879.

Talley, Robert. "Newton J. Jones Visits Here, Remembers 1878 Plague Well." *The Commercial Appeal,* July 19, 1938.

Taubes, Gary. "Tales of a Bloodsucker—Asian Tiger Mosquitoes." *Discover* (July 1998).

Thomas, Hugh. *Cuba, or the Pursuit of Freedom.* New York: Da Capo Press, 1998.

Thornton, Charles. "Yellow Fever's Horror Recalled 100 Years After Its Departure." *Press-Scimitar,* August 7, 1978.

Truby, Albert E. *Memoir of Walter Reed: The Yellow Fever Episode.* New York: Paul B. Hoeber, 1943.

Turner, Charles. *The Celebrant.* Ann Arbor, MI: Servant Publications, 1982.

Van Epps, Heather L. "Broadening the Horizons for Yellow Fever: New Uses for an Old Vaccine." *Journal of Experimental Medicine,* Vol. 201, No. 1: 165–68.

Vaughan, Victor Clarence. *A Doctor's Memories.* Indianapolis: Bobbs-Merrill, 1926.

Victory, Joy. "Rare U.S. Case of Yellow Fever Ends in Death." *Corpus Christi Caller-Times,* March 27, 2002.

Waring, George E. *Report on the Condition of the Sewers of Memphis, Tenn.* March 4, 1893.

Waring, George E. *Report on the Social Statistics of Cities.* Washington, D.C., 1887.

Warner, Margaret H. "Hunting the Yellow Fever Germ: The Principle and Practice of Etiological Proof in Late Nineteenth-Century America." *Bulletin of Historical Medicine* 59 (1985): 361–82.

Watts, Sheldon. *Epidemics and History: Disease, Power and Imperialism.* New Haven and London: Yale University Press, 1997.

White, Mimi. "1878 Yellow Fever Epidemic." *Tennessee Medical Alumnus,* Vol. II, No. 2 (Fall 1978).

Williams, Greer. *The Plague Killers.* New York: Charles Scribner's Sons, 1969.

Williams, Greer. *The Virus Hunters.* New York: Alfred A. Knopf, 1960.

Wills, Christopher. *Yellow Fever, Black Goddess: The Coevolution of People and Plagues.* Cambridge: Helix Books, Perseus Publishing, 1996.

Wingfield, Marshall. "The Life and Letters of Dr. William J. Armstrong." *West Tennessee Historical Society Papers,* Vol. IV (1950): 97–113.

Winter, F. "The Romantic Side of the Conquest of Yellow Fever." *The Military Surgeon,* Vol. 61 (1927).

Wood, Laura. *Walter Reed, Doctor in Uniform.* New York: Julian Messner, 1943.

World Health Organization. *Prevention and Control of Yellow Fever in Africa.* 1998.

World Health Organization. *Strengthening Global Preparedness for Defense against Infectious Disease Threats, Senate Hearing on The Threat of Bioterrorism and the Spread of Infectious Diseases.* September 5, 2001.

"Yellow Fever." *Old Shelby County Magazine,* No. 5 (1999).

"The Yellow Fever Epidemic in Memphis in 1878." Supplement to *The West Tennessee Catholic.*

"The Yellow Fever Experiments in Cuba." *Journal of American Medical Association* 37 (1901): 839–40.

"Yellow Fever in New Orleans." *The New York Times,* July 26, 1878.

Newspaper Clippings

The Avalanche, 1878
The Commercial Appeal, 1970–2005
Frank Leslie's Illustrated Newspaper, 1878

The Memphis Daily Appeal, 1878–1879
The New York Times, 1878–1879, 1900–1901
Press-Scimitar, 1878–1978
The Washington Post, 1900–1901

Websites

Britannica On-Line Encyclopedia
The Centers for Disease Control and Prevention
Federal Research Division, Library of Congress
Global Alliance for Vaccines and Immunizations
The National Institutes of Health, Library of Medicine
Nobel Prize Foundation
Pan-American Health Organization
The Philip S. Hench Walter Reed On-Line Collection
The Rockefeller Foundation
The Rutherford B. Hayes Presidential Center On-Line Diary
The World Health Organization